Student Solutions Manual

TWELFTH EDITION
Modern Elementary Statistics

John E. Freund
Benjamin M. Perles

PEARSON

Prentice Hall

Upper Saddle River, NJ 07458

Editor-in-Chief: Sally Yagan
Executive Acquisitions Editor: Petra Recter
Supplement Editor: Joanne Wendelken
Executive Managing Editor: Kathleen Schiaparelli
Assistant Managing Editor: Karen Bosch
Production Editor: Jennifer Zisa
Supplement Cover Manager: Paul Gourhan
Supplement Cover Designer: Christopher Kossa
Manufacturing Buyer: Ilene Kahn
Manufacturing Manager: Alexis Heydt-Long

© 2007 Pearson Education, Inc.
Pearson Prentice Hall
Pearson Education, Inc.
Upper Saddle River, NJ 07458

The author and publisher of this book have used their best efforts in preparing this book. These efforts include the development, research, and testing of the theories and programs to determine their effectiveness. The author and publisher make no warranty of any kind, expressed or implied, with regard to these programs or the documentation contained in this book. The author and publisher shall not be liable in any event for incidental or consequential damages in connection with, or arising out of, the furnishing, performance, or use of these programs.

Printed in the United States of America

10 9 8 7 6 5 4 3 2 1

ISBN 0-13-187442-X

Pearson Education Ltd., *London*
Pearson Education Australia Pty. Ltd., *Sydney*
Pearson Education Singapore, Pte. Ltd.
Pearson Education North Asia Ltd., *Hong Kong*
Pearson Education Canada, Inc., *Toronto*
Pearson Educación de Mexico, S.A. de C.V.
Pearson Education—Japan, *Tokyo*
Pearson Education Malaysia, Pte. Ltd.

Table of Contents

CHAPTER

1

Introduction

1.1 **a.** The results can be misleading because "Xerox copiers" is often used as a generic term for photocopiers.

   b. Since Rolex watches are very expensive, persons wearing them can hardly be described as average individuals.

   c. The cost of such cruises are greater than the cost of a typical vacation.

1.3 **a.** Many persons are reluctant to give honest answers about their personal health habits.

   b. Successful graduates are more likely to return the questionnaire than graduates that have not done so well.

1.5 **a.** Since $4 + 2 = 6$ and $3 + 3 = 6$, the statement is descriptive.

   b. The data relate to a given day, so that "always" requires a generalization.

   c. The data relate to a given day, so that a statement about what happens over a week requires a generalization.

   d. Since the statement does not tell us anything about the time required to treat a patient of either sex, the statement requires a generalization.

1.7 **a.** The statement is a generalization based upon the misconception that trucks necessarily get better mileage on rural roads.

b. The statement is a generalization based on the idea that higher speeds lead to poorer mileage.

c. Since 15.5 occurs twice while each of the other figures occurs only once, the statement is purely descriptive.

d. Since none of the values exceeds 16.0, the statement is purely descriptive.

1.9 **(a)** The conclusion is nonsense. The same number of elevators go up and down.

   (b) When operating, most elevators in this building will typically be on the upper floors; and few elevators will be on the first and second floors. Thus, the first elevator to stop on the third floor is likely to be coming down from above.

1.11 Nominal

1.13 **a.** Interval

   b. Ordinal

   c. Ratio

1

2.1

b.

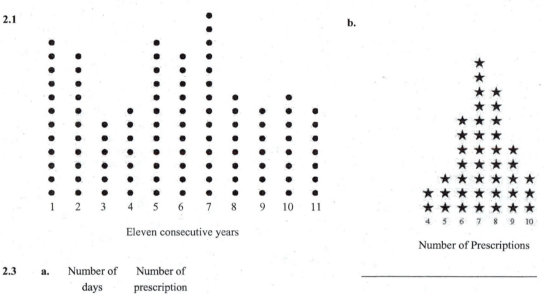

Eleven consecutive years

Number of Prescriptions

2.3 **a.**

Number of days	Number of prescription
4	2
5	3
6	7
7	11
8	9
9	5
10	3
	40

2.5

Afghan ● ● ● ● ●
Basset ● ●
Beagle ● ● ● ● ● ● ● ●
Bloodhound ●
Dachshund ● ● ● ● ● ● ● ●
Greyhound ● ● ● ● ● ●

2.7 A ○ ○ ○ ○ ○ ○ ○

B ○ ○ ○ ○ ○

C ○ ○ ○ ○

D ○ ○

E ○

2

2.9

Codes

3 ● ● ● ● ● ● ● ● ● ● ● ● ● ● ● ●

2 ● ● ● ● ● ● ● ● ●

0 ● ● ● ● ● ●

1 ● ● ●

4 ● ●

2.11   **a.** 36, 31, 37, 35, and 32

        b. 415, 438, 450, and 477

        c. 254, 254, 250, 253, and 259

2.13

5	8	6							
6	5	6	4	0	7				
7	9	7	8	1	2	1	3	5	
8	6	4	3	8	1	1	5	9	0
9	5								

2.15

16.	6														
17.	3	0													
18.	4	9	1	3	3	2	6	5	6						
19.	2	6	3	5	0	4	4	8	6	7	5	8	4	3	5 8
20.	4	4	2	1	3	7	3	8	4	2	9	5	7	6	1
21.	8	0	4	5	5	7	9	1							
22.	9	7													
23.	5														

2.17

6	55	75	32							
7	84	83	60	60	18					
8	34	65	39	88	31	86	42	54	26	66	65
9	19	12	39	61	54	01				

2.19

1.3	7								
1.4	2	4	6	9					
1.5	0	2	3	3	4	4	8	8	9
1.6	0	2	3	6	8				
1.7	2								

2.21

8	4	8								
9	2	3	6	7	7	9				
10	1	3	3	3	4	5	5	6	8	9
11	0	3	5							
12	2	4	7							

2.23   A convenient choice would be 220–239, 240–259, 260–279, 280–299, 300–319, 320–339, 340–359, 360–379.

2.25   **a.** 0–49.99, 50.00–99.99, 100.00–149.99, 150.00–199.99

        b. 20.00–49.99, 50.00–79.99, 80.00–109.99, 110.00–139.99, 140.00–169.99, 170.00–199.99

        c. 30.00–49.99, 50.00–69.99, 70.00–89.99, 90.00–109.99, 110.00–129.99, 130.00–149.99, 150.00–169.99, 170.00–189.99

2.27   **a.** 5.0, 20.0, 35.0, 50.0, 65.0, and 80.0

        b. 19.9, 34.9, 49.9, 64.9, 79.9, and 94.9

        c. 4.95, 19.95, 34.95, 49.95, 64.95, 79.95, and 94.95

        d. 15

2.29   There is no provision for values from 50.00 to 59.99, and values from 70.00 to 79.99 go into two classes.

2.31   There is no provision, for example, for cookies or jello. Also, there is ambiguity about classifying, say, fruit cake, pie and ice cream, fruit with ice cream, etc.

2.33   **a.** 20–24, 25–29, 30–34, 35–39, 40–44

   b. 22, 27, 32, 37, and 42

   c. All 5's

2.35   **a.** 60.0–74.9, 75.0–89.9, 90.0–104.9, 105.0–119.9, and 120.0–134.9

   b. 67.45, 82.45, 97.45, 112.45, and 127.45

2.37   The respective percentages are 2.5, 5.0, 37.5, 40.0, 10.0, and 5.0 percent.

2.39   The respective class frequencies are 13, 14, 16, 12, 4, and 1.

2.41   The cumulative percentages corresponding to 19 or less, 24 or less, 29 or less, 34 or less, 39 or less, 44 or less, and 49 or less, are, respectively, 0, 21.67, 45.0, 71.67, 91.67, 98.33, and 100.00 percent.

2.43   The cumulative class frequencies corresponding to more than 0.49, more than 0.59, ..., and more than 149 are 120, 118, 112, 100, 62, 36, 23, 16, 8, 3, and 0.

2.45   The cumulative percentages corresponding to 20 or more, 25 or more, ..., and 55 or more are, respectively, 100, 93.75, 79.17, 56.25, 31.25, 14.58, 6.25, 0.

2.47   Student project.

2.49

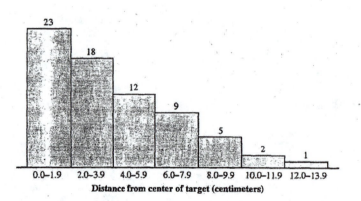

4

2.51   Various possibilities

2.53   a.

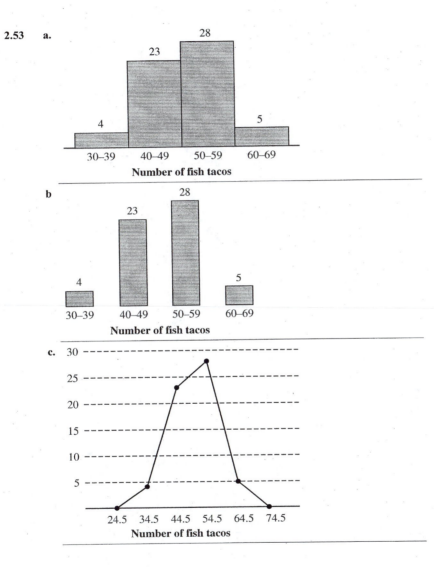

5

d

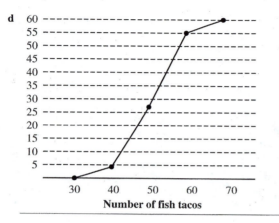

Number of fish tacos

2.55   The cumulative frequencies corresponding to less than 0.20, less than 0.40, less than 0.60, less than 0.80, less than 1.00, less than 1.20, less than 1.40, and less than 1.60 are 0, 3, 16, 42, 62, 72, 79, and 80.

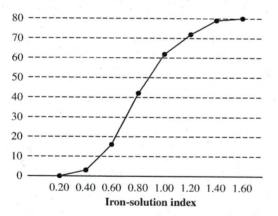

Iron-solution index

2.57   It might easily give a misleading impression because we tend to compare the areas of rectangles rather than their heights. Since the 80–99 class is twice as wide as the others, we could make the areas of the four rectangles proportional to the class frequencies by dividing the height of the 80–99 rectangle by 2.

2.59   The central angles corresponding to the eight classes are, $\frac{1,586}{5,179} \cdot 360° = 110.2°$, $\frac{805}{5,179} \cdot 360° = 56.0°$,

$\frac{761}{5,179} \cdot 360° = 52.9°$, $\frac{598}{5,179} \cdot 360° = 41.6°$, $\frac{393}{5,179} \cdot 360° = 27.3°$, $\frac{301}{5,179} \cdot 360° = 20.9°$, $\frac{267}{5,179} \cdot 360° = 18.6°$, and $\dfrac{468}{5,179} \cdot 360° = 32.5°.$

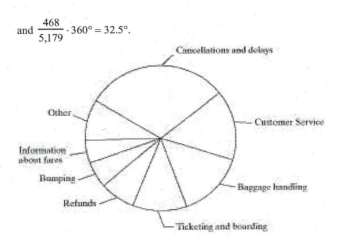

2.61   Student project.

2.63   The frequencies corresponding to the five categories are 4, 11, 24, 9, and 2, and the corresponding central angles are 28.8, 79.2, 172.8, 64.8, and 14.4 degrees.

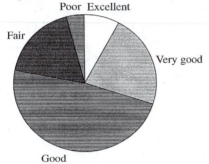

2.65   The Exercise requires the use of a computer or of a graphing calculator.

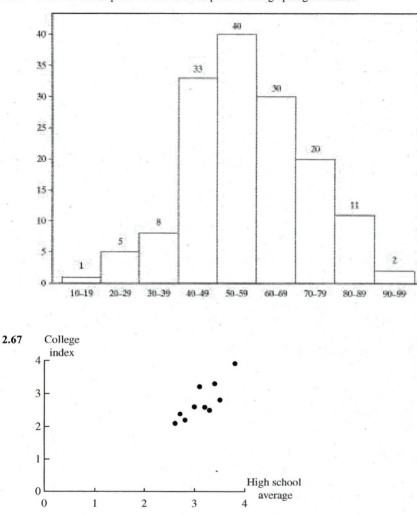

2.67   College

There is an upward linear trend, but the points are fairly widely scattered.

CHAPTER 3

Summarizing Data:
Measures of Location

3.1 **a.** If we are interested only in the blood pressures of patients in this specific cardiac ward, the data constitute a population.

b. If we want to generalize about patients in other cardiac wards, or in other cardiac patients, the data constitute a sample.

3.3 The information would be looked upon as a sample if it is to be used for planning future tournaments. The information would be looked upon as a population if it is to be used to pay off the tennis club's employees who were to receive a bonus for each day there was a rain delay.

3.5
$$\bar{x} = \frac{112 + 83 + 102 + 84 + 105 + 121 + 76 + 110 + 98 + \ldots + 85}{12}$$
$$= \frac{1,170}{12}$$
$$= 97.5$$

3.7
$$\bar{x} = \frac{9.96 + 9.98 + 9.92 + 9.98 + 9.96}{5} = \frac{49.80}{5} = 9.96 \text{ ml.}$$
On the average, the calibration is off by $10 - 9.96 = 0.04$ ml.

3.9 The total weight of the 18 persons is $18 \cdot 166 = 2,988$ pounds. Since 2,988 does not exceed 3,200, there is no danger of the elevator being overloaded.

3.11 $\bar{x} = \dfrac{7 + 6 + 7 + 0 + 7 + 9 + 6 + 0}{8} = 5.25.$ It is at best a conjecture that this relatively low figure accounts for relatively poor performance.

3.13 $\dfrac{(32)(78) + (48)(84)}{32 + 48} = \dfrac{2,496 + 4,032}{80} = 82$ points

3.15 **a.** At most $\dfrac{33.5}{50} = 0.67.$

b. At most $\dfrac{17.2}{20} = 0.86.$

3.17 **a.** The square root of $9 \cdot 36 = \sqrt{324} = 18$

b. The fourth root of $1 \cdot 2 \cdot 8 \cdot 81 = 16 \cdot 81$ is $2 \cdot 3 = 6.$

c. The geometric mean of the two growth rates is the square root of $\dfrac{3}{2} \cdot \dfrac{8}{3} = 4$ is 2. The predictions for the fourth and fifth days are, respectively, $48 \cdot 2 = 96$ and $96 \cdot 2 = 192.$

3.19 $\bar{x}_w = \dfrac{6,000(0.0375)+10,000(0.0396)+4,000(0.0325)}{20,000} = 0.03755$. The total return on the 3 investments is

225 + 396 + 130 = 751. This is $\dfrac{751}{20,000} = 0.03755$, which equals the weighted mean of the percentages.

3.21 $\bar{x}_w = \dfrac{382(33,373)+(450)(31,684)+113(40,329)}{382 + 450 + 113} = \dfrac{31,563,463}{945} = \$33,400.49.$

3.23 $\bar{x} = 78.27$ minutes.

3.25   **a.** Since $\dfrac{55+1}{2} = 28$, the median is the 28th value.

        b. Since $\dfrac{34+1}{2} = 17.5$, the median is the mean of the 17th and 18th values.

3.27   Arranged according to size, the data are 38, 40, 40, 50, 53, 53, 57, 59, 63, 65, 66, and 68. Since $\dfrac{12+1}{2} = 6.5$,

the median is the mean of the 6th and 7th values, namely, $\dfrac{53+57}{2} = 55$.

3.29   Arranged according to size, the data are 113, 117, 121, 122, 126, 128, 130, 133, 134, 135, 137, 138, 139, 140, 140, 142, 142, 142, 143, 145, 146, 147, 148, 150, 151, 155, 157, 157, 158, 159, 164, and 169. Since $\dfrac{32+1}{2} = 16.5$, the median is the mean of the 16th and 17th values, namely, $\dfrac{142+142}{2} = 142$ minutes.

3.31   Arranged according to size, the original data are 225, 238, 265, 332, 340, and 346, and their median is $\dfrac{265+332}{2} = 298.5$. With 238 replaced by 832, the data are 225, 265, 332, 340, 346, and 832, and their

median is $\dfrac{332+340}{2} = 336$, so that the error is only 336 − 298.5 = 37.5.

3.33

8	2 7
9	2 5 5
10	0 1 2 3 4 4 5 6 6 8
11	0 0 1 3 3 3 4 5 5 6 7 7 8 8 8 9 9 9
12	0 0 1 3 4 5 5 5 6 6 6 6 7 8 9 9
13	2 2 3 5 6 7 7
14	3 6 6 8

Since $\dfrac{60+1}{2} = 30.5$, the median is the mean of the 30th and 31st values, namely, $\dfrac{118+119}{2} = 118.5$ grams.

3.37   Since the three midranges are 29.8, 30.0, and 30.3, the manufacturers of car C can use the midrange to substantiate the claim that their car performed best.

3.39   **a.** Since $\dfrac{41+1}{2} = 21$, the median is the 21st value. Since $\dfrac{20+1}{2} = 10.5$, Q_1 is the mean of the 10th and 11th values, and Q_3 is the mean of the 10th and 11th values from the other end.

        b. Since $\dfrac{50+1}{2} = 25.5$, the median is the mean of the 25th and 26th values. Since $\dfrac{25+1}{2} = 13$, Q_1 is the 13th value and Q_3 is the 13th value from the other end.

3.41   Since $\dfrac{34+1}{2} = 17.5$, the median is the mean of the 17th and 18th values. Since $\dfrac{17+1}{2} = 9$, Q_1 is the 9th value and Q_3 is the 9th value from the other end. There are eight values to the left of the Q_1 position, eight values between the Q_1 position and the median position, eight values between the median position and the Q_3 position, and eight values to the right of the Q_3 position.

3.43   The smallest value is 41 and the largest value is 66. Also, from Exercise 3.42, $Q_1 = 47$, the median is 56, $Q_3 = 62$, so that the boxplot is

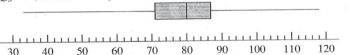

3.45   The smallest value is 405 and the largest value is 440. Also, from Exercise 3.44, $Q_1 = 411$, the median is 417, and $Q_3 = 432$, so that the boxplot is

3.47   The smallest value is 33 and the largest value is 118. Also, from Exercise 3.46, $Q_1 = 71$, the median is 80, and $Q_3 = 87$, so that the boxplot is

3.49   The smallest value is 82 and the largest value is 148. From Exercise 3.48, $Q_1 = 109$, the median is 118.5, and $Q_3 = 126.5$.

3.51   The mode is 48.

3.53   The mode is 0, which occurs six times. There seems to be a cyclical (up and down) pattern, which does not follow by just giving the mode.

3.55   Occasionally is the mode.

3.57   **a.** The mean and the median can both be determined.

        b. The mean cannot be determined because of the open class; the median can be determined because it does not fall into one of the open classes.

    c. The mean cannot be determined because of the open class; the median cannot be determined because it falls into the open class.

3.59 The mean is 4.88 and the median is 4.89, both rounded to two decimal places.

3.61 The mean is 47.64 and the median is 46.20, both rounded to two decimal places.

3.63 Since P_{95} would have fallen into the open class, it could not have been determined.

3.65 $Q_1 = 0.82$, the median is 0.90, the mean is 0.94, and $Q_3 = 1.04$, all rounded to two decimal places.

3.67   **a.** $\sum_{i=1}^{5} z_i$

      b. $\sum_{i=5}^{12} x_i$

      c. $\sum_{i=1}^{6} x_i f_i$

      d. $\sum_{i=1}^{3} y_i^2$

      e. $\sum_{i=1}^{7} 2x_i$

      f. $\sum_{i=2}^{4} (x_i - y_i)$

      g. $\sum_{i=2}^{5} (z_i + 3)$

      h. $\sum_{i=1}^{4} x_i y_i f_i$

3.69   **a.** $2 + 3 + 4 + 5 + 6 = 20$

      b. $2 + 8 + 9 + 3 + 2 = 24$

      c. $4 + 24 + 36 + 15 + 12 = 91$

      d. $8 + 72 + 144 + 75 + 72 = 371$

3.71   **a.** 10, 6, 1, and 13

      b. 8, 12, and 10

3.73 $\left(\sum_{i=1}^{2} x_i\right)^2 = (x_1 + x_2)^2 = x_1^2 + 2x_1 x_2 + x_2^2$

12

$$\sum_{i=1}^{2} x_i^2 = x_1^2 + x_2^2$$

It is not a true statement.

CHAPTER

Summarizing Data:
Measures of Variation

4.1   **a.** The range is 2.70–2.63 = 0.07

     b.

x	$x - \bar{x}$	$(x - \bar{x})^2$
2.64	-0.02	0.0004
2.70	+0.04	0.0016
2.67	+0.01	0.0001
2.63	-0.03	0.0009
10.64		0.0030

For use in table, $\bar{x} = 2.66$.

$$s = \sqrt{\frac{\Sigma(x - \bar{x})^2}{n-1}} = \sqrt{\frac{0.0030}{4-1}} = 0.032 \text{ rounded to 3 decimal places.}$$

4.3   **a.** The range is 23 – 12 = 11

     b. $\Sigma x = 17 + 20 + 12 + 14 + 18 + 23 + 17 + 19 + 18 + 15 = 173$

$$\Sigma x^2 = 289 + 400 + 144 + 196 + 324 + 529 + 289 + 361 + 324 + 225 = 3081$$

$$S_{xx} = 3081 - \frac{(173)^2}{10} = 88.10$$

$$S = \sqrt{\frac{S_{xx}}{n-1}} = \sqrt{\frac{88.10}{9}} = \sqrt{9.7889} = 3.1287. \text{ Rounding to two decimal places, 3.13.}$$

4.5   Arraying the data of Exercise 4.3 we get the following marked with the location and values of the quartiles and median.

          12   14   15   17   17   18   18   19   20   23

                       Q_1 Median Q_3

Twice the interquartile range is 2(19 – 15) = 8. The range in Exercise 4.3 is 11. It is not a surprise that the range is greater than twice the interquartile range because many types of data tend to cluster around the median.

4.7   2.3452, or 2.3 rounded to one decimal.

4.9

x	$x - \mu$	$(x - \mu)^2$
7	-1.5	2.25
8	-0.5	0.25
4	-4.5	20.25
11	+2.5	6.25
13	+4.5	20.25
15	+6.5	42.25
6	-2.5	6.25
4	-4.5	20.25
68	0.0	118.00

$$\mu = \frac{68}{8} = 8.5$$

$$\sigma = \sqrt{\frac{\Sigma(x-\mu)^2}{N}} = \sqrt{\frac{118.00}{8}} \approx 3.84$$

This Exercise can also be solved and the same answer obtained by using the computing formula.

4.11 The range is $8.98 - 8.92 = 0.06$ for part (a); and $0.08 - 0.02 = 0.06$ for part (b). The range was not affected by the subtraction of the constant value.

4.13   **a.** Subtracting the smallest value from the largest value we get $14 - 4 = 10$ claims.

b.

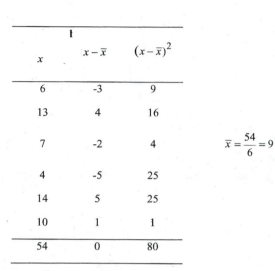

x	$x - \bar{x}$	$(x - \bar{x})^2$
6	-3	9
13	4	16
7	-2	4
4	-5	25
14	5	25
10	1	1
54	0	80

$$\bar{x} = \frac{54}{6} = 9$$

$$s = \sqrt{\frac{80}{6-1}} = 4.0$$

c. First we calculate Σx and Σx^2. $\Sigma x = 6 + 13 + 7 + 4 + 14 + 10 = 54$; and then $\Sigma x^2 = 36 + 169 + 49 + 16 + 196 + 100 = 566$. Then, substituting these totals and $n = 6$ into the formula for S_{xx} and into the formula for s we get

$$s = \sqrt{\frac{80}{6-1}} = 4.0$$

4.15 Verify

4.17 $s = 20.68$

4.19 **a.** $1 - \dfrac{1}{6.25} = 0.84$ or 84%

       b. $1 - \dfrac{1}{256} = \dfrac{255}{256}$ or 0.996 which is 99.6%

4.21 **a.** Between 94.8 and 128.4 minutes

       b. Between 83.6 and 139.6 minutes

4.23 The percentages are 65%, 97.5%, and 100%, which are close to 68%, 95%, and 99.7%.

4.25

A	B
$x_A = 76.75$	$x_B = 49.50$
$\bar{x}_A = 58.25$	$\bar{x}_B = 37.50$
$s_A = 11.0$	$s_B = 4.00$
$z_A = \dfrac{76.75 - 58.25}{11.00}$	$z_B = \dfrac{49.50 - 37.50}{4.00}$
$= 1.68$	$= 3.00$

$3.00 > 1.68.$ Stock B is relatively more overpriced.

4.27

Brass	Rainfall
$\Sigma x = 260$	$\Sigma x = 0.80$
$\Sigma x^2 = 13,540$	$\Sigma x^2 = 0.1640$
$S_{xx} = 20$	$S_{xx} = 0.0040$
$s = 2.236$	$s = 0.0365$
$\bar{x} = 52$	$\bar{x} = 0.20$
$V = \dfrac{2,236}{52} \cdot 100 = 4.3\%$	$V = \dfrac{0.0365}{0.20} \cdot 100 = 18.25\%$

Since 18.25% is greater than 4.3%, the rainfall data are relatively more variable.

4.29   An alternative measure of relative variation is the coefficient of quartile variation which is defined as

$$\frac{Q_3 - Q_1}{Q_3 + Q_1} \cdot 100$$

Arraying the time in seconds data of Exercise 4.8 we get

$$76 \quad 83 \quad 84 \;\bigg|\; 85 \quad 91 \quad 98 \;\bigg|\; 102 \quad 103 \quad 105 \;\bigg|\; 110 \quad 112 \quad 121$$

$$Q_1 = 84.5 \qquad Q_2 = 100 \qquad Q_3 = 107.5$$

$$\frac{107.5 - 84.5}{107.5 + 84.5} \cdot 100 = 12\%, \text{ rounded to whole numbers}$$

4.31   **a.** Temperatures in degrees Fahrenheit and time in seconds cannot be compared directly because they are expressed in different units of measure. The answers do not tell us whether they reflect a great deal of variation or very little variation.

b. Using the coefficients of quartile variation as a measure of relative variation, the data of Exercise 4.8 are relatively more variable than those of Exercise 4.11.

4.33 Following is the distribution of the percentages of students (from Exercise 3.58) in 50 elementary schools who are bilingual.

Percentages	Frequency	x Class marks	xf	x^2f
0 – 4	18	2	36	72
5 – 9	15	7	105	735
10 – 14	9	12	108	1,296
15 – 19	7	17	119	2,023
20 – 24	1	22	22	484
	$\Sigma f = 50$		$\Sigma xf = 390$	$\Sigma x^2f = 4,610$

$$S_{xx} = 4,610 - \frac{(390)^2}{50} = 1,568$$

$$s = \sqrt{\frac{1,568}{49}} = 5.66$$

4.35 This is the grouped distribution of the compressive strength in (1,000 psi) of 120 samples of concrete:

Compressive strength	Frequency	Class Mark x	xf	x^2f
4.20 – 4.39	6	4.295	25.770	110.682
4.40 – 4.59	12	4.495	53.940	242.460
4.60 – 4.79	23	4.695	107.985	506.990
4.80 – 4.99	40	4.895	195.800	958.441
5.00 – 5.19	24	5.095	122.280	623.017
5.20 – 5.39	11	5.295	58.245	308.407
5.40 – 5.59	4	5.495	21.980	120.780
	$\Sigma f = 120$		$\Sigma xf = 586.000$	2,870.777

$$S_{xx} = 2{,}870.777 - \frac{(586.000)^2}{120} = 9144$$

$$s = \sqrt{\frac{9.144}{119}} = 0.277$$

4.37

Grades	f Number of Students	x Class Marks	xf	$x^2 f$
$10-24$	44	17	748	12,716
$25-39$	70	32	2,240	71,680
$40-54$	92	47	4,324	203,228
$55-69$	147	62	9,114	565,068
$70-84$	115	77	8,855	681,835
$85-99$	32	92	2,944	270,848
	500	$\Sigma\, xf =$ 28,225		$\Sigma\, x^2 f = 1{,}805{,}375$

a. $\bar{x} = \dfrac{28{,}225}{500} = 56.45$

$\tilde{x} = 54.5 + \dfrac{44}{147} \cdot 15 = 58.99$

b. $S_{xx} = 1{,}805{,}375 - \dfrac{(28{,}225)^2}{500}$

$= 1{,}805{,}375 - 1{,}593{,}301.25$

$= 212{,}073.75$

$s = \sqrt{\dfrac{212{,}073.75}{499}} = 20.615$

$= 20.62$

4.39 $SK = \dfrac{3(50.167 - 50.571)}{7.448}$

$= -0.16$

19

4.41   The smallest value is 2 and the largest value is 35, $Q_1 = 11$, the median is 23.4, the mean is 20.8. The data are negatively skewed, which gives a negative value of SK.

4.43   Smallest value 3.3

            Largest value 6.0

            Median 4.25

            Q_1 . . . . 3.7

            Q_3 . . . . 4.6

            The long tail on the right suggests that the data are positively skewed.

4.45   The frequencies corresponding to 0, 1, 2, 3, and 4 are 27, 17, 4, 1, and 1. The distribution is reverse J-shaped and positively skewed.

4.47   The frequencies corresponding to 0 through 5 are, respectively, 23, 11, 5, 3, 2, and 16. This distribution is U-shaped.

Review Exercises for Chapters 1, 2, 3, and 4

R.1   The numbers 23 and 24 can go into the third and fourth classes; the distribution does not accommodate the numbers 36, 37, 38, and 39.

R.3   The measurements are 123, 125, 130, 134, 137, 138, 141, 143, 144, 146, 146, 149, 150, 152, 152, 155, 158, 161, and 167.

R.4   Arranged according to size, the data are 3.9, 4.6, 4.9, 5.3, 5.8, 6.3, 6.3, 6.5, 6.8, 7.3, 7.4, 7.5, 7.7, 8.2, 8.5, 9.0,

R.5

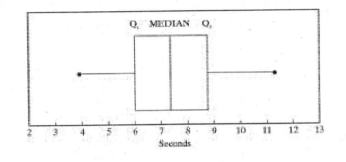

R.7

Class Mark	Frequency		
x	f	fx	fx^2
2	34	68	136
7	20	140	980
12	15	180	2,160
17	9	153	2,601
22	2	44	968
	80	$\Sigma fx = 585$	$\Sigma fx^2 = 6.845$

a. The mean is 7.31 rounded to two decimal places.

b. The median is 6.

c. Standard deviation is 5.70.

d. $SK = \dfrac{3(7.31 - 6)}{5.70} = 0.69$

R.9   **a.**

0.04	5 6 7 8 9 9
0.05	0 2 2 4 4 4 5 5 6 7 7 8 8 8
0.06	0 1 2 2 3 3 5 6 7 8
0.07	2 2

      b. The median is 0.057. Q_1 is 0.052 and Q_3 is 0.0625.

      The data are slightly positively skewed.

R.11   **a.** The data would constitute a population if the meteorologist is interested only in the given ten years.

      b. The data would constitute a sample if the meteorologist is interested in making predictions for future years.

R.13   The two coefficients of variation are, respectively, $V = \dfrac{s}{\bar{x}} \cdot 100 = \dfrac{2}{10} \cdot 100 = 20\%$ for the first company; and

      $\dfrac{3}{25} \cdot 100 \approx 12\%$ for the other company. There is relatively less variability in the firm where V is smaller $(12\% < 20\%)$.

R.15   At least $\left(1 - \dfrac{1}{3^2}\right) \cdot 100 = 88.9\%$ (rounded to one decimal) have diameters between 23.91 and 24.09 mm.

R.17   **a.** $9 + 2 = 11$

      b. $20 + 15 + 9 = 44$

      c. cannot be determined

      d. cannot be determined

R.19 $\Sigma\, fx^2 = 755.$ $\bar{x} = \dfrac{755}{60} = 12.58$

      Location of the median is $\dfrac{60}{2} = 30$th value.   Median $= 9.5 + \dfrac{13}{23} \cdot 5 = 12.33$

      $\Sigma fx = 11{,}205$

      $S_{xx} = 11{,}205 - \dfrac{(755)^2}{60} = 1{,}704.5833$

      $s = \sqrt{\dfrac{1{,}704.5833}{59}} = 5.38$

R.21   **(a)**

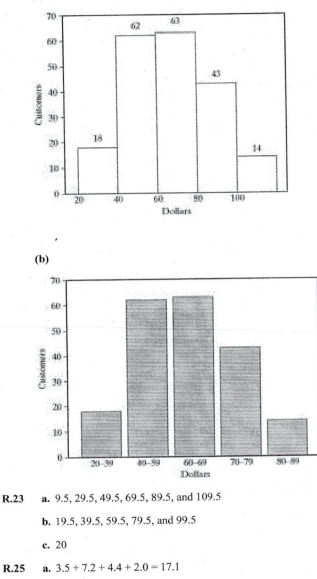

(b)

R.23   **a.** 9.5, 29.5, 49.5, 69.5, 89.5, and 109.5

       b. 19.5, 39.5, 59.5, 79.5, and 99.5

       c. 20

R.25   **a.** $3.5 + 7.2 + 4.4 + 2.0 = 17.1$

       b. $3.5^2 + 7.2^2 + 6.4^2 + 2.0^2 = 87.45$

       c. $17.1^2 = 292.41$

R.27 $\dfrac{s}{19.5} \cdot 100 = 32$, so that s = 6.24.

23

R.29   There are other kinds of fibers and also shirts made of combinations of fibers.

R.31   **a.** Cannot be determined

       b. Yes, the number in the fourth class

       c. Yes, the sum of the numbers in the second and third classes

       d. Cannot be determined

R.33   **a.** Referring to the practice as "unfair" is begging the question.

       b. A difference in opinion between persons having telephones and persons not having telephones may affect the results.

R.35   **a.** 14.5, 29.5, 44.5, 59.5, 74.5, 89.5, 109.5, and 119.5.

       b. 22, 37, 52, 67, 82, 97, and 112.

       c. 15.

R.37   The cumulative "less than" frequencies are 0, 3, 17, 35, 61, 81, 93, and 100.

R.39   12   4

       13   0 0 5

       14   2 6 9

       15   1 3 4 5 6 8 9

       16   2 2 2 5

       17   2 3

       18   2

       19

       20   4

R.41 $57 - 47 = 47 - 37 = 10 = 2.5k$, $k = 4$, and the percentage is at least $\left(1 - \dfrac{1}{4^2}\right) \cdot 100\% = 93.75\%$.

R.43   0   ☆ ☆ ☆ ☆ ☆ ☆ ☆ ☆ ☆ ☆ ☆

       1   ☆ ☆ ☆ ☆ ☆ ☆ ☆ ☆ ☆

       2   ☆ ☆ ☆ ☆ ☆

       3   ☆ ☆ ☆

       4   ☆

       5   ☆

R.45 $V = \dfrac{0.537}{2.1} \cdot 100 \approx 25.57\%$.

R.47   It is assumed that the difference between A and B counts for as much as the differences between B and C, the difference between C and D, and the difference between D and E.

CHAPTER 5

Possibilities
and
Probabilities

5.1 Possible Monday and Tuesday sales of 0 and 0, 0 and 1, 0 and 2, 1 and 0, 1 and 1, 2 and 0, 2 and 1, and 2 and 2.

5.3 American League team wins 5th game and the series; American League team loses 5th game, wins 6th game and the series; American League team loses 5th game, loses 6th game, wins 7th game and series; National League team wins 5th game, 6th game, wins 7th game and series.

5.5 $50 \times 49 \times 48 \times 47 = 5,527,200$

5.7 $6! = 6 \cdot 5 \cdot 4 \cdot 3 \cdot 2 \cdot 1 = 720$

$5! = 5 \cdot 4 \cdot 3 \cdot 2 \cdot 1 = 120$

$4! = 4 \cdot 3 \cdot 2 \cdot 1 = 24$

$3! = 3 \cdot 2 \cdot 1 = 6$

5.9 $12^P4 = 12 \cdot 11 \cdot 10 \cdot 9 = 11,880$

5.11 **a.** In three cases.

b. In two cases.

5.13 **a.** 0 and 0, 0 and 1, 0 and 2, 1 and 0, 1 and 1, and 2 and 0.

b. Label the paintings Q and R, and let N denote none. The possibilities are N and N, N and Q, N and R, N and Q and R, Q and N, Q and R, R and N, R and Q, Q and R and N.

5.15 $6 \cdot 4 = 24$

5.17 $4 \cdot 32 = 128$

5.19 **a.** 4

b. $4 \cdot 4 = 16$

c. $4 \cdot 3 = 12$

5.21 $5 \cdot 5 \cdot 2 \cdot 3 = 150$

5.23 **a.** $3^{10} = 59,049$

b. $2^{10} = 1,024$

5.25 **a.** True

b. False

c. $3! + 0! = 6 + 1 = 7$, true

d. $6! + 3! = 720 + 6 \neq 362,880$; false

e. $\dfrac{9!}{7!2!} = \dfrac{9 \cdot 8}{2} = 36$; true

f. $17! = 17 \cdot 16 \cdot 15! \neq 15! \cdot 2$; false

5.27 $6! = 720$

5.29 $\dfrac{32!}{5!27!} = \dfrac{32 \cdot 31 \cdot 30 \cdot 29 \cdot 28}{120} = 201,376$

5.31 $8 \cdot 7 \cdot 6 \cdot 5 \cdot 4 = 6,720$

5.33 $9! = 362,880$

5.35 **a.** $4 \cdot 3 \cdot 3 \cdot 2 = 72$

b. $3 \cdot 3 \cdot 2 = 18$

c. $3 \cdot 3 = 9$

5.37 **a.** $\dfrac{7!}{2! \cdot 2!} = 1,260$

b. $\dfrac{6!}{3! \cdot 3!} = 20$

c. $\dfrac{6!}{2! \cdot 2! \cdot 2!} = 90$.

5.39 $\dfrac{15 \cdot 14 \cdot 13}{6} = 455$

26

5.41 $\dfrac{18 \cdot 17 \cdot 16}{6} = 816$

5.43 **a.** $\dfrac{1}{52}$

  b. $\dfrac{6}{52} = \dfrac{3}{26}$

  c. $\dfrac{12}{52} = \dfrac{3}{13}$

  d. $\dfrac{13}{52} = \dfrac{1}{4}$

5.45 $s = 4 \cdot 4 \cdot 4, n = \dfrac{52 \cdot 51 \cdot 50}{3 \cdot 2} = 22,100$

  and $\dfrac{s}{n} = \dfrac{4 \cdot 4 \cdot 4}{22,100} = \dfrac{16}{5,525}$

5.47 The 36 possible outcomes are 1 and 1, 1 and 2, 1 and 3, 1 and 4, 1 and 5, 1 and 6, 2 and 1, 2 and 2, 2 and 3, 2 and 4, 2 and 5, 2 and 6, 3 and 1, 3 and 2, 3 and 3, 3 and 4, 3 and 5, 3 and 6, 4 and 1, 4 and 2, 4 and 3, 4 and 4, 4 and 5, 4 and 6, 5 and 1, 5 and 2, 5 and 3, 5 and 4, 5 and 5, 5 and 6, 6 and 1, 6 and 2, 6 and 3, 6 and 4, 6 and 5, 6 and 5, and 6 and 6.

  a. $\dfrac{3}{36} = \dfrac{1}{12}$

  b. $\dfrac{4}{36} = \dfrac{1}{9}$

  c. $\dfrac{8}{36} = \dfrac{2}{9}$

5.49 **a.** $\dfrac{15}{72} = \dfrac{5}{24}$

  b. $\dfrac{50}{72} = \dfrac{25}{36}$

  c. $\dfrac{7}{72}$

  d. $\dfrac{35}{72}$

5.51 **a.** $\dfrac{37}{75}$

  b. $\dfrac{15}{75} = \dfrac{1}{5}$

  c. $\dfrac{16}{75}$

5.53 **a.** $\dfrac{56}{220} = \dfrac{14}{55}$

  b. $\dfrac{48}{220} = \dfrac{12}{55}$

5.55 **a.** $\dfrac{475}{1,683}$

  b. $\dfrac{96}{462}$

5.57 $\dfrac{424}{954} = \dfrac{4}{9}$

5.59 $\dfrac{678}{904} = \dfrac{3}{4}$

5.61 $\dfrac{28}{52} = \dfrac{7}{13}$

5.65 By the same token we could argue that there is no life elsewhere in the universe, that there is only plant life elsewhere in the universe, that there is only animal life elsewhere in the universe, or that there are both kinds of life forms in the universe. Having no information whatsoever, the principle of equal ignorance would lead to a probability of $\dfrac{1}{4}$ that there is no life elsewhere in the universe.

5.69 The additional information can go either way so the probability can increase or decrease.

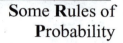

Some Rules of Probability

6.1 **a.** $U' = \{a, c, d, f, g\}$, the scholarship is awarded to Ms. Adam, Miss Clark, Mrs. Daly, Ms. Fuentes, or Ms. Gardner. $U \cap V = \{e, h\}$; the scholarship is awarded to Mr. Earl or Mr. Hall. $U \cup V' = \{a, b, c, d, e, h\}$; the scholarship is not awarded to Ms. Fuentes or Ms. Gardner.

6.3 **a.** $\{(0, 2), (1, 1), (2, 0)\}$

      b. $\{(0, 0), (1, 1)\}$

      c. $\{(1, 1), (2, 1), (1, 2)\}$

6.5 **a.** There is one professor less than there are assistants.

      b. Altogether there are four professors and assistants.

      c. There are two assistants.

K and L are mutually exclusive; K and M are not mutually exclusive; and L and M are not mutually exclusive.

6.7 **a.** $\{(4, 1), (3, 2)\}$

      b. $\{(4, 3)\}$

      c. $\{(3, 3), (4, 3)\}$

6.9 **a.** $K' = \{(0, 0), (1, 0), (2, 0), (3, 0), (0, 1), (1, 1), (2, 1)\}$; at most one boat is rented out for the day.

      b. $L \cap M = \{(2, 1), (3, 0)\}$

6.11 **a.** $\{A, D\}$

      b. $\{C, E\}$

      c. $\{B\}$

6.13 **a.** Not mutually exclusive; there can be sunshine and rain on the same day.

      b. Not mutually exclusive; why not?

      c. Mutually exclusive; when it is 11 P.M. in Los Angeles it is already the next day in New York.

      d. Not mutually exclusive; one could be a bachelor's degree and the other could be a master's degree.

6.15 **a.** $98 - 50 = 48$

      b. $224 - 50 = 174$

      c. $360 - 224 - 48 = 88$

6.17 **a.** The car needs an engine overhaul, transmission repairs, and new tires.

      b. The car needs transmission repairs and new tires, but no engine overhaul.

      c. The car needs an engine overhaul, but no transmission repairs and no new tires.

d. The car needs an engine overhaul and new tires.

e. The car needs transmission repairs but no new tires.

f. The car does not need transmission repairs.

6.19   **a.** $8 + 8 + 5 + 3 = 24$

       b. $3 + 8 + 3 + 2 = 16$

6.21 $P(C')$ is the probability that there will not be enough capital for the planned expansion. $P(E')$ is the probability that the planned expansion will not provide enough parking. $P(C' \cap E)$ is the probability that there will not be enough capital for the planned expansion and that the planned expansion will provide enough parking. $P(C \cap E')$ is the probability that there will be enough capital for the planned expansion but that the planned expansion will not provide enough parking.

6.23 $P(A')$ is the probability that the attendance at the concert will not be good. $P(A' \cup W)$ is the probability that the attendance will not be good and/or more than half the crowd will walk out during the intermission. $P(A \cap W)$ is the probability that there will be a good attendance at the concert and at most half the crowd will walk out during the intermission.

6.25   **a.** Postulate 1

       b. Postulate 2

       c. Postulate 2

       d. Postulate 3

6.27   The corresponding probabilities are $\dfrac{2}{2+1} = \dfrac{2}{3}$ and $\dfrac{3}{3+1} = \dfrac{3}{4}$, and since $\dfrac{2}{3} + \dfrac{3}{4} > 1$, the odds cannot be right.

6.29   **a.** Since A is contained in $A \cup B$, $P(A)$ cannot exceed $P(A \cup B)$.

       b. Since $A \cap B$ is contained in A, $P(A \cap B)$ cannot exceed $P(A)$.

6.31   If $P(A) = 0$

6.33   **a.** The odds for getting at least two heads in four flips of a balanced coin are 11 to 5.

       b. The probability is $\dfrac{34}{55}$ that at least one of the tiles will have a blemish.

       c. The odds are 19 to 5 that any particular household will not be included.

       d. The probability is $\dfrac{719}{720}$ that not all the letters will end up in the right envelopes.

6.35   The probability is greater than or equal to $\dfrac{6}{11}$ but less than $\dfrac{3}{5}$.

6.37   The probability for the \$1,000 raise is $\dfrac{5}{12}$, the probability for the \$2,000 raise is $\dfrac{1}{12}$, and the probability for either raise is $\dfrac{1}{2}$. Since $\dfrac{5}{12} + \dfrac{1}{12} = \dfrac{1}{2}$, the probabilities are consistent.

6.39 $\dfrac{a}{b} = \dfrac{p}{1-p}$ yields $a(1 - p) = bp$, $a - ap = bp$, $a = ap + bp$, $a = p(a + b)$, and $p = \dfrac{a}{a+b}$.

6.41 $1 - (0.19 + 0.26 + 0.25 + 0.20 + 0.07)$ $= 0.03$

6.43   **a.** $0.23 + 0.15 = 0.38$

       b. $0.31 + 0.24 + 0.07 = 0.62$

       c. $0.23 + 0.24 = 0.47$

       d. $1 - 0.07 = 0.93$

6.45   The probabilities are $\dfrac{1}{32}, \dfrac{5}{32}, \dfrac{10}{32}, \dfrac{10}{32}, \dfrac{5}{32}$, and $\dfrac{1}{32}$.

6.47 $0.33 + 0.27 - 0.19 = 0.41$

6.49 $0.39 + 0.46 - 0.31 = 0.54$

6.51   **a.** $P(A|T)$

     b. $P(W|A)$

     c. $P(T|W')$

     d. $P(W|A' \cap T')$

6.53   **a.** $P(M|I)$

     b. $P(I'|A')$

     c. $P(I' \cap A'|N)$

6.55 $\dfrac{1}{3} = \dfrac{0.2}{0.6}$

6.57 $\dfrac{3}{7} = \dfrac{0.3}{0.7}$

6.59 $\dfrac{0.44}{0.80} = 0.55$

6.61 $\dfrac{\binom{30}{2}}{\binom{40}{2}} = \dfrac{29}{52}$

6.63   Since $(0.80)(0.95) = 0.76$; the two events are independent.

6.65   0.42, 0.18, and 0.12.

6.67 $(0.25)(0.40)^2(0.60) = 0.024$

6.69 $(0.70)(0.70)(0.30)(0.60)(0.40) = 0.03528$

6.71 $\dfrac{(0.40)(0.66)}{0.372} = 0.71$ rounded to two decimal places.

6.73 $\dfrac{(0.50)(0.68)}{0.76} = 0.447$ rounded to three decimals.

6.75 $\dfrac{(0.10)(0.95)}{(0.10)(0.95) + (0.90)(0.05)} = 0.679$ rounded to three decimal places.

6.77   **a.** $\dfrac{3}{4} \cdot \dfrac{1}{3} + \dfrac{1}{4} \cdot \dfrac{3}{4} = \dfrac{7}{16}$

     b. $\dfrac{\frac{3}{16}}{\frac{7}{16}} = \dfrac{3}{7}$

6.79   The probabilities for the respective causes are 0.229, 0.244, 0.183, and 0.344 all rounded to three decimals. On the basis of this information, the most likely cause is purposeful action.

CHAPTER

Expectations and Decisions

7.1 $750 \cdot \dfrac{1}{3,000} = \$0.25$

7.3 $\dfrac{3,000 + 1,000}{15,000} = \$0.27$ rounded up to the nearest cent.

7.5
a. $E = 300,000 \cdot \dfrac{1}{2} + 120,000 \cdot \dfrac{1}{2} = \$210,000$

Each golfer has the expectation of $210,000.

b. If the younger golfer is favored by odds of 3 to 2, his probability of winning is $\dfrac{3}{5}$.

E younger =
$\dfrac{3}{5}(300,000) + \dfrac{2}{5}(120,000) = \$228,000.$

E older
$= \dfrac{2}{5}(300,000) + \dfrac{3}{5}(120,000) = \$192,000.$

7.7 $16,000(0.25) + 13,000(0.46) + 12,000(0.19) + 10,000(0.10) = \$13,260$, so that the expected gross profit is $13,260 − $12,000 = $1,260.

7.9 $7,500 > (30,000)p$

$\dfrac{7,500}{30,000} > p$

$0.25 > p$

7.11 $50,000 − 12,500p < 45,000$, so that $5,000 < 12,500p$ and $p > \dfrac{5,000}{12,500} = 0.40.$

7.13 $x \cdot \dfrac{1}{2} = 1,000 \cdot \dfrac{1}{2} + 400$, so that $\dfrac{x}{2} = 900$ and $x = 2 \cdot 900 = \$1,800.$

7.15 The two expectations are
$120,000 \cdot \dfrac{3}{4} - 30,000 \cdot \dfrac{1}{4} = \$82,500$ and
$180,000 \cdot \dfrac{1}{2} - 45,000 \cdot \dfrac{1}{2} = \$67,500$; the contractor should take the first job.

7.17 If the driver goes to the barn first, the expected distance is
$(18 + 18) \cdot \dfrac{1}{6} + (18 + 8 + 22) \cdot \dfrac{5}{6} = 46$ miles. If the driver goes to the shopping center first, the expected distance is
$(22 + 22) \cdot \dfrac{5}{6} + (22 + 8 + 18) \cdot \dfrac{1}{6} = 44\dfrac{2}{3}$ miles.
He should go first to the shopping center.

7.19 If the driver goes to the barn first, the expected distance is
$(18 + 18) \cdot \dfrac{1}{4} + (18 + 8 + 22) \cdot \dfrac{3}{4}$
$= 45$ miles; if the driver goes to the shopping center first, the expected distance is
$(22 + 22) \cdot \dfrac{3}{4} + (22 + 8 + 18) \cdot \dfrac{1}{4}$
$= 45$ miles. It does not matter where he goes first.

7.21 If they continue, the expected profit is − $300,000; if they do not continue, the expected profit is −$300,000. It does not matter whether or not they continue the operation.

7.23
a. The maximum losses would be $600,000 if the tests are continued and $500,000 if the tests are discontinued. To minimize the maximum loss, the tests should be discontinued.

b. If he first goes to the barn, the possible distances are 36 and 48 miles, and if he first goes to the shopping center, the possible distances are 44 and 48 miles. In either case, the maximum distance is 48 miles, so it does not matter where he goes first.

7.25   **a.** The maximum profit would be
$4,500,000 if the operation is continued
and $450,000 if it is not continued. Thus,
the maximum profit would be maximized
if the operation is continued.

b. The worst that can happen is a loss of
$2,700,000 if the operation is continued
and a loss of $1,800,000 if it is not
continued. Therefore, the worst that can
happen is minimized if the operation is
discontinued.

7.27   **a.** The errors are 0, 1, 4, or 5, and
correspondingly, the consultant will get
600, 580, 280, or 100 dollars. He can
expect to get

$$600 \cdot \frac{2}{5} + 580 \cdot \frac{1}{5} + 280 \cdot \frac{1}{5} + 100 \cdot \frac{1}{5}$$
$$= \$432.$$

b. The errors are 1, 0, 3, or 4, and
correspondingly, the consultant will get
$580, $600, $420, or $280 dollars. He
can expect to get

$$580 \cdot \frac{2}{5} + 600 \cdot \frac{1}{5} + 420 \cdot \frac{1}{5} + 280 \cdot \frac{1}{5}$$
$$= \$492.$$

7.29   **a.** The median, 18

b. The mean, 19

Review Exercises for Chapters 5, 6, and 7

R.49   **a.** $P(C) = 0.12 + 0.48 = 0.60$

       b. $P(D') = 0.12 + 0.08 = 0.20$

       c. $P(C \cup D) = 0.92$

       d. $P(C \cap D') = 0.12$

R.51   **a.** $0.01 + 0.02 + 0.05 + 0.14 + 0.16 = 0.38$

       b. $0.18 + 0.15 + 0.09 = 0.42$

       c. $0.14 + 0.16 + 0.20 = 0.50$

R.53

R.55 $\dfrac{3}{4} \le p < \dfrac{4}{5}$

R.57 $\dbinom{15}{4} = 1,365$

R.59 $\dfrac{1,134}{1,800} = 0.63$

R.61   **a.** If the mortgage manager accepts or rejects the application, the expected profits are, respectively, $8,000(0.9) - 20,000(0.1) = 5,200$ and 0. To maximize the expected profit, the mortgage manager should accept the application.

       b. If the mortgage manager accepts or rejects the application, the expected profits are, respectively, $8,000(0.7) - 20,000(0.3) = -400$ and 0. To maximize the expected profit, the mor

mortgage manager should reject the application.

       c. If the mortgage manager accepts or rejects the application, the maximum losses are, respectively, 20,000 and 0. To minimize the maximum loss, the mortgage manager should reject the application.

R.63 $2^8 \cdot 4^4 = 65,536$

R.65   **a.** $11 \cdot 15 = 165$

       b. $12 \cdot 14 = 168$

R.67 $(0.24)^3 = 0.014$ rounded to three decimals.

R.69   **a.** The probability is $\dfrac{3}{21+3} = \dfrac{1}{8}$ that the driver will win.

       b. The probability is $\dfrac{11}{11+5} = \dfrac{11}{16}$ that at most two of the cards will be black.

R.71   **a.** Since $P(A \cap B) = 0.62 - (0.37 + 0.25) = 0$, events A and B are mutually exclusive.

       b. Since $(0.37)(0.25) \ne 0$, events A and B are not independent.

       c. $(0.60)(0.15) = 0.09 \ne 0.075$

R.73 $30 = 3 \cdot \dfrac{15}{60} + V \cdot \dfrac{45}{60}$. Simplification yields

$120 = 3 + 3V$, and $V = \$39$.

R.75 **a.** The accountant should use the mode, 30.

    b. The accountant should use the mean, 31.

R.77 **a.** $5! = 120$

    b. $\dfrac{6!}{2!} = 360$

    c. $\dfrac{6!}{2!2!} = 180$

    d. $\dfrac{7!}{3!} = 840$

R.79 **a.** $\dfrac{\binom{3}{1}\binom{2}{1}\binom{5}{1}}{\binom{10}{3}} = \dfrac{1}{4}$

    b. $\dfrac{\binom{5}{3}}{120} = \dfrac{1}{12}$

    c. $\dfrac{\binom{3}{1}\binom{5}{2}}{120} = \dfrac{1}{4}$

R.81 **a.** $\dfrac{(0.02)(0.90)}{(0.02)(0.90) + (0.98)(0.08)} = \dfrac{0.0180}{0.0964}$

$= 0.187$

    rounded to three decimals

    b. $\dfrac{(0.98)(0.92)}{(0.02)(0.90) + (0.98)(0.92)} = \dfrac{0.9016}{0.9196}$

$= 0.980$

    rounded to three decimals

R.83 There are many persons who would prefer a guaranteed 4.5% to a potentially risky 6.2%.

R.85 There are 12 outcomes in event A, and 8 outcomes in the event outside A. If p is the probability of each outcome outside A, then

$12 \cdot 2p + 8p = 1$ and $p = \dfrac{1}{32}$. Therefore,

$$P(A) = 12 \cdot \dfrac{2}{32} = \dfrac{24}{32} = \dfrac{3}{4}.$$

R.87 $\dfrac{\binom{18}{10}}{2^{18}} = \dfrac{43{,}758}{262{,}144} = 0.167$ rounded to three decimal places.

R.89 **a.** $4! = 24$

    b. $5 \cdot 4! = 120$

CHAPTER 8

8.1   **a.** No; $0.52 + 0.26 + 0.32 = 1.10 > 1.00$

   b. No; $0.18 + 0.02 + 1.00 = 1.20 > 1.00$

   c. Yes; the values are all non-negative and their sum equals 1.

8.3   **a.** Yes; the values are all non-negative and $7 \cdot \frac{1}{7} = 1$

   b. No; the sum of the values is $10 \cdot \frac{1}{9} > 1$.

   c. Yes; the values are all non-negative and $\frac{3}{18} + \frac{4}{18} + \frac{5}{18} + \frac{6}{18} = 1$.

8.5 $\binom{3}{2}\left(\frac{3}{4}\right)^2\left(\frac{1}{4}\right) = \frac{27}{64} = 0.42$ rounded to two decimal places.

8.7 $\binom{4}{0}(0.10)^0(0.90)^4 = 0.6561$; the value in Table V is 0.656 rounded to three decimals.

8.9   **a.** $0.028 + 0.121 + 0.233 + 0.267 = 0.649$

   b. $0.037 + 0.009 + 0.001 = 0.047$

8.11   **a.** $0.002 + 0.007 + 0.024 = 0.033$

   b. $0.177 + 0.207 + 0.186 + 0.127 = 0.697$

   c. $0.127 + 0.063 + 0.022 + 0.005 = 0.217$

8.13   **a.** $1 - 0.282 = 0.718$

   b. 0.069

   c. 0.014

8.15   **a.** $0.0000 + 0.0008 + 0.0063 + 0.0285 + 0.0849 = 0.1205$

   b. 0.1205

8.17   **a.** $(0.40)(0.60)^3 = 0.0864$

   b. $(0.25)(0.75)^4 = 0.079$ rounded to three decimal places

   c. $(0.70)(0.30)^2 = 0.063$

35

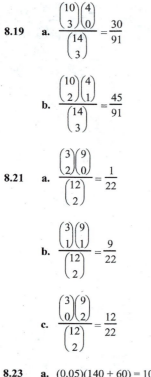

8.19　**a.**　$\dfrac{\dbinom{10}{3}\dbinom{4}{0}}{\dbinom{14}{3}} = \dfrac{30}{91}$

b.　$\dfrac{\dbinom{10}{2}\dbinom{4}{1}}{\dbinom{14}{3}} = \dfrac{45}{91}$

8.21　**a.**　$\dfrac{\dbinom{3}{2}\dbinom{9}{0}}{\dbinom{12}{2}} = \dfrac{1}{22}$

b.　$\dfrac{\dbinom{3}{1}\dbinom{9}{1}}{\dbinom{12}{2}} = \dfrac{9}{22}$

c.　$\dfrac{\dbinom{3}{0}\dbinom{9}{2}}{\dbinom{12}{2}} = \dfrac{12}{22}$

8.23　**a.**　$(0.05)(140 + 60) = 10$
　　　　Since $n = 12 > 10$, condition is not
　　　　satisfied.

b.　$(0.05)(220 + 280) = 25$
　　　Since $n = 20 < 25$, condition is satisfied.

c.　$(0.05)(250 + 390) = 32$
　　　Since $n = 30 < 32$, condition is satisfied.

d.　$(0.05)(220 + 220) = 22$
　　　Since $n = 25 > 22$, condition is not
　　　satisfied.

8.25　The binomial approximation is 0.0750
　　　rounded to four decimal places. Since the
　　　hypergeometric probability is 0.0762, the
　　　error of the binomial approximation is
　　　$0.0750 - 0.0762 = 0.0012$ rounded to four
　　　decimal places.

8.27　**a.**　Since $np = 12.5 > 10$, conditions are not
　　　　satisfied.

b.　Since $n = 400 > 100$ and $np = 8 < 10$, the
　　　conditions are satisfied.

c.　Since $n = 90 < 100$, the conditions are
　　　not satisfied.

8.29　$f(3) = \dfrac{6^3(0.002479)}{3!} = 0.089$　rounded to
　　　three decimal places.

8.31　$np = 150(0.05) = 7.5$
　　　$\dfrac{7.5^0 \cdot e^{-7.5}}{0!} = 0.00055$　rounded to five

　　　decimal places, $\dfrac{7.5^1 e^{-7.5}}{1!} = 0.00415$, and

　　　$\dfrac{7.5^2 e^{-7.5}}{2!} = 0.01555$

　　　rounded to five decimal places. The
　　　probability for at most two will be involved
　　　in an accident is
　　　$0.00055 + 0.00415 + 0.01555 = 0.02025$
　　　rounded to five decimals.

8.33　Since $n = 120 < 0.05(3,200) = 160$, the given
　　　hypergeometric distribution can be
　　　approximated with the binomial distribution
　　　with $n = 120$ and $p = \dfrac{50}{3,200} = 0.0156$. Then
　　　since $n = 120 > 100$ and
　　　$120(0.0156) = 1.87 < 10$, this binomial
　　　distribution can be approximated with the
　　　Poisson distribution with $np = 1.87$.

8.35　**a.**　$\dfrac{1.6^0 \cdot e^{-1.6}}{0!} = 0.2019$

b.　$\dfrac{1.6^1 \cdot e^{-1.6}}{1!} = 0.3230$

c.　$\dfrac{1.6^2 \cdot e^{-1.6}}{2!} = 0.2584$
　　　all rounded to four decimal places.

8.37　$\dfrac{10!}{6!\,3!\,1!}(0.70)^6(0.20)^3(0.10)^1 = 0.0791$
　　　rounded to four decimal places.

8.39　$\dfrac{10!}{7!\,1!\,1!\,1!}(0.60)^7(0.20)(0.10)(0.10) = 0.0403$
　　　rounded to four decimal places.

8.41 $\sigma^2 = \sum x^2 f(x) - \mu^2$

$\sigma^2 = 1^2(0.4) + 2^2(0.3) + 3^2(0.2) + 4^2(0.1) - 2^2$

$\sigma^2 = 0.4 + 1.2 + 1.8 + 1.6 - 4.0$

$\sigma^2 = 5.0 - 4.0 = 1.0$

8.43 $\mu = 0(0.0035) + 1(0.0231) + 2(0.0725) + \cdots + 12(0.0004) = 4.8587$

$\sigma^2 = 0^2(0.0035) + 1^2(0.0231) + 2^2(0.0725) + \cdots + 12^2(0.0004) - (4.8587)^2$
     $= 3.5461$ rounded to four decimal places.

$\sigma = \sqrt{3.5461} = 1.883$ rounded to three decimal places.

8.45 $\mu = 4 \cdot \dfrac{1}{2} = 2$, $\sigma^2 = 4 \cdot \dfrac{1}{2} \cdot \dfrac{1}{2} = 1$, and $\sigma = 1$.

8.47   **a.** $\mu = 484 \cdot \dfrac{1}{2} = 242$, $\sigma^2 = 484 \cdot \dfrac{1}{2} \cdot \dfrac{1}{2} = 121$, and $\sigma = 11$.

       b. $\mu = 120$ and $\sigma = 10$

       c. $\mu = 180$ and $\sigma = 11.225$ rounded to three decimal places.

       d. $\mu = 24$, and $\sigma = 4.8$.

       e. $\mu = 520$, and $\sigma = 13.491$ rounded to three decimals.

8.49 $\mu = 0(0.013) + 1(0.128) + 2(0.359) + 3(0.359)$
         $+ 4(0.128) + 5(0.013) = 2.5$

$\mu = \dfrac{8 \cdot 5}{5 + 11} = \dfrac{40}{16} = 2.5$

8.51 $\mu = 0(0.082) + 1(0.205) + 2(0.256) + \cdots + 9(0.001)$

      $= 2.501$, which is very close to $\lambda = 2.5$.

8.53   **a.** The proportion of heads will be between 0.475 and 0.525;

       b. Same as (a) with value of n changed.

9.1 **a.** First area is bigger.

   b. First area is bigger.

   c. Areas are equal.

9.3 **a.** First area is bigger.

   b. Second area is bigger.

   c. Areas are equal.

9.5 **a.** $0.1064 + 0.5000 = 0.6064$.

   b. $0.5000 + 0.4032 = 0.9032$

   c. $0.5000 - 0.2852 = 0.2148$

9.7 **a.** $2(0.2794) = 0.5588$.

   b. $0.4177 + 0.5000 = 0.9177$.

9.9 **a.** $1.63 = z$.

   b. $0.9868 - 0.5000 = 0.4868$

9.11 **a.** $0.4678 - 0.3023 = 0.1655$;

   b. $0.4772 - 0.2157 = 0.2615$.

9.13 **a.** $z = 2.03$

   b. $z = 0.98$

   c. $z = \pm 1.47$

   d. $z = -0.41$

9.15 **a.** $2(0.3413) = 0.6826$

   b. $2(0.4772) = 0.9544$

   c. $2(0.4987) = 0.9974$

   d. $2(0.49997) = 0.99994$

   e. $2(0.4999997) = 0.9999994$

9.17 **a.** 0.9332

   b. 0.7734

   c. 0.2957

   d. 0.9198

9.19 Since the entry in Table I closest to
$0.5000 - 0.2000 = 0.3000$ is 0.2995
corresponding to $z = 0.84$,
$\dfrac{79.2 - 62.4}{\sigma} = 0.84$ and $\sigma = \dfrac{16.8}{0.84} = 20$.

9.21 **a.** $1 - e^{-0.4} = 0.3297$

   b. $(1 - e^{-0.9}) - (1 - e^{-0.5})$
   $= 0.1999$

9.23 **a.** $e^{-2} = 0.1353$

   b. $e^{-3} = 0.049787$

   c. $1 - e^{-0.5} = 0.3935$

9.25 **a.** $z = \dfrac{19.0 - 17.4}{2.2} = 0.73$; the normal curve
   area is 0.2678 and the probability is 0.77.

   b. $z = \dfrac{12.0 - 17.4}{2.2} = -2.45$ and

   $z = \dfrac{15.0 - 17.4}{2.2} = -1.09$; the normal
   curve areas are 0.4930 and 0.3621, and
   the probability is 0.13.

9.27 **a.** $z = \dfrac{33.4 - 38.6}{6.5} = -0.80$ the normal
   curve area is 0.2881 and the probability
   is 0.7881.

   b. $z = \dfrac{34.7 - 38.6}{6.5} = 0.60$; the normal curve
   area is 0.2257 and the probability is
   $0.5000 - 0.2257 = 0.2743$

9.29 $\dfrac{x-18.2}{1.2} = -0.84,\ x = 17.2$ ounces.

9.31 $z = \dfrac{2.0-1.96}{0.08} = 0.50$; the normal curve area

is 0.1915 and the percentage is

$(0.5000 - 0.1915)\cdot 100 = 30.85\%$

9.33   The entry corresponding to

$z = \dfrac{79.0-63}{20} = 0.80$ is 0.2881 and

$(0.5000 - 0.2881)8{,}000{,}000 = 1{,}695{,}200$
or approximately 1.7 million men would
be discharged.

9.35   **a.** $\dfrac{18-25.8}{\sigma} = -0.84$, so that

$\sigma = \dfrac{7.8}{0.84} = 9.3$

b. $z = \dfrac{30-25.8}{9.3} = 0.45$; the normal curve

area is 0.1736 and the probability is
$0.5000 - 0.1736 = 0.33.$

9.37 $\dfrac{400-338}{\sigma} = 1.556,$ so that

$\sigma = \dfrac{62}{1.556} = 39.8;$

$z = \dfrac{300-338}{39.8} = -0.955$ and the probability
is $0.5000 - 0.3302 = 0.17.$

9.39 $\mu = 10$ and $\sigma = \sqrt{5} = 2.236;$

$z = \dfrac{11.5-10}{2.236} = 0.67$

a.   probability is $0.500 - 0.2486 = 0.2514;$

b.   probability $= 0.120 + 0.074 + 0.037 +$
$0.015 + 0.005 + 0.001 = 0.252.$

9.41   **a.** $np = 32\cdot\dfrac{1}{7} = 4.57$; conditions not
satisfied;

b. $np = 75(0.10) = 7.5$ and $n(1-p) = 67.5$;
conditions satisfied;

c. $np = 50(0.08) = 4$; conditions not
satisfied.

9.43 $\mu = 200(0.80) = 160$ and $\sigma = \sqrt{32} = 5.66$;
$z = \dfrac{169.5-160}{5.66} = 1.68$ and the
probability is

9.45   **a.**   Using Fig. 9.24 and normal curve
approximation the probabilities are
0.6261 and 0.6141; error is
$0.6261 - 0.6141 = 0.0120$ and percentage
error is $\dfrac{0.0120}{0.6261}\cdot 100 = 1.9\%.$

b.   Using Fig. 9.24 and normal curve
approximation the probabilities are
0.1623 and 0.1639; error is
$0.1639 - 0.1623 = 0.0016$ and percentage
error is $\dfrac{0.0016}{0.1623}\cdot 100 = 0.99\%.$

9.47 $\mu = 100(0.18) = 18$

$\sigma = \sqrt{100(0.18)(0.82)} = 3.84$

$z = \dfrac{11.5-18}{3.84} = -1.69$

The probability is $0.4545 + 0.5000 =$
$0.9545.$

CHAPTER

Sampling and Sampling Distributions

10.1   **a.** $\binom{6}{2} = 15$

      b. $\binom{20}{2} = 190$

      c. $\binom{32}{2} = 496$

      d. $\binom{75}{2} = 2{,}775$

10.3   **a.** $\dfrac{1}{\binom{12}{4}} = \dfrac{1}{495}$

      b. $\dfrac{1}{\binom{20}{4}} = \dfrac{1}{4{,}845}$

10.5   *uvw, uvx, uvy, uvz, uwx, uwy, uwz, uxy, uxz, uyz, vwx, vwy, vwz, vxy, vxz, vyz, wxy, wxz, wyz, xyz*

10.7 $\dfrac{4}{20} = \dfrac{1}{5}$

10.9   **a.** $\dfrac{1}{\binom{5}{3}} = \dfrac{1}{10}$

      b. $\dfrac{\binom{4}{2}}{10} = \dfrac{6}{10} = \dfrac{3}{5}$

      c. $\dfrac{\binom{3}{1}}{10} = \dfrac{3}{10}$

10.11   3406, 3591, 3383, 3554, 3513, 3439, 3707, 3416, 3795, and 3329.

10.13   264, 429, 437, 419, 418, 252, 326, 443, 410, 472, 446, and 318.

10.15   6094, 2749, 0160, 0081, 0662, 5676, 6441, 6356, 2269, 4341, 0922, 6762, 5454, 7323, 1522, 1615, 4363, 3019, 3743, 5173, 5186, 4030, 0276, 7845, 5025, 0792, 0903, 5667, 4814, 3676, 1435, 5552, 7885, 1186, 6769, 5006, 0165, 1380, 0831, 3327, 0279, 7607, 3231, 5015, 4909, 6100, 0633, 6299, 3350, 3597.

10.17 $\dfrac{3}{5} \cdot \dfrac{2}{4} \cdot \dfrac{1}{3} = \dfrac{1}{10}$

10.19 $\dfrac{n}{N} \cdot \dfrac{n-1}{N-1} \cdot \dots \cdot \dfrac{1}{N-n+1}$

$= \dfrac{n}{N} \cdot \dfrac{n-1}{N-1} \cdot \dots \cdot \dfrac{1}{N-n+1} \cdot \dfrac{(N-n)!}{(N-n)!}$

$= \dfrac{n!(N-n)!}{N!}$

$= \dfrac{1}{\dbinom{N}{n}}$

10.21 16.8, 24.0, 20.1, 21.9, 15.8, 22.1, 20.9, 21.3, 18.8, and 18.5;
27.6, 15.9, 24.2, 15.2, 20.4, 21.1, 15.7, 25.0, 16.9, and 25.0;
25.4, 16.9, 19.4, 17.2, 19.0, 25.8, 16.8, 12.9, 21.1, and 13.2;
16.6, 19.4, 16.6, 16.1, 16.9, 18.0, 20.5, 15.0, 27.9, and 16.7;
22.6, 17.0, 17.0, 18.6, 18.4, 15.5, 17.0, 15.8, 14.7, and 24.2

10.23 All the December figures that are, of course, much higher than the others, go into the same (sixth) sample.

10.25 **a.** $\dbinom{9}{2}\dbinom{3}{2} = 36 \cdot 3 = 108$

b. $\dbinom{9}{3}\dbinom{3}{1} = 84 \cdot 3 = 252$

10.27 $n_1 = \dfrac{250}{1,000} \cdot 40 = 10, \; n_2 = \dfrac{600}{1,000} \cdot 40 = 24,$

$n_3 = \dfrac{100}{1,000} \cdot 40 = 4, \quad n_4 = \dfrac{50}{1,000} \cdot 40 = 2.$
and

10.29 **a.** $n_1 = \dfrac{100 \cdot 10,000 \cdot 45}{10,000 \cdot 45 + 30,000 \cdot 60} = 20,$ and $n_2 = 100 - 20 = 80.$

b. $n = \dfrac{84 \cdot 5,000 \cdot 15}{5,000 \cdot 15 + 2,000 \cdot 18 + 3,000 \cdot 5}$

$= 50$

$n_2 = \dfrac{84 \cdot 2,000 \cdot 18}{126,000} = 24,$ and

$n_3 = 84 - (50 + 24) = 10.$

10.31 **a.** $\dfrac{4+5+4}{25} = \dfrac{13}{25} = 0.52$

b. $\dfrac{3+4+5+4+3}{25} = \dfrac{19}{25} = 0.76$

10.33   **a.** Finite, consisting of the twenty applicants.

            b. Infinite, consisting of the hypothetically infinite possible number of weighings.

            c. Infinite, consisting of the hypothetically infinite possible number of flips of a coin.

            d. Finite, consisting of 25 cards.

            e. Infinite, consisting of the hypothetically infinite possible number of determinations of mileage

10.35   **a.** It is divided by $\sqrt{\dfrac{120}{30}} = 2$

            b. Multiplied by $\sqrt{\dfrac{245}{5}} = 7$

10.37   **a.** $\sqrt{\dfrac{90}{99}} = 0.954$ rounded to three decimal places

            b. $\sqrt{\dfrac{275}{299}} = 0.958$ rounded to three decimal places

            c. $\sqrt{\dfrac{4,900}{4,999}} = 0.990$ rounded to three decimal places

10.39 $0.6745 \cdot \dfrac{12.8}{\sqrt{60}} = 1.115$

10.41 $\sigma_{\bar{x}} = \dfrac{2.4}{\sqrt{25}} = 0.48$

            a. Since $k = \dfrac{1.2}{0.48} = 2.5$, the probability is at least $1 - \dfrac{1}{2.5^2} = 0.84$.

            b. Since $z = 2.50$, the probability is $2(0.4938) = 0.9876$.

10.43 $\sigma_{\bar{x}} = \dfrac{0.025}{\sqrt{16}} = 0.00625$

            $z = \dfrac{0.01}{0.00625} = 1.6$ and the probability is $2(0.4452) = 0.8904$.

10.45 $\dfrac{\sigma}{\sqrt{144}} = 1.25 \cdot \dfrac{\sigma}{\sqrt{n}}$, so that $\sqrt{n} = 12(1.25) = 15$ and $n = 225$.

10.47   The medians are 11, 15, 19, 17, 14, 13, 14, 17, 18, 14, 14, 19, 16, 18, 17, 18, 19, 17, 14, 12, 11, 16, 16, 19, 18, 17, 17, 15, 13, 14, 15, 16, 14, 15, 12, 16, 16, 17, 14, and 17. Their standard deviation is 2.20 rounded to two decimal places and the corresponding standard error formula yields $1.25 \cdot \dfrac{4}{\sqrt{5}} = 2.24$ rounded to two decimal places.

10.49 The sum of the forty sample variances is 615.9, and their mean is $\frac{615.9}{40} = 15.40$ rounded to two decimal places and the percentage error is $\frac{16 - 15.40}{16} \cdot 100 = 3.75\%$.

Review Exercises for Chapters 8, 9, and 10

R.91   **a.** $\dfrac{\binom{10}{4}}{\binom{18}{4}} = \dfrac{210}{3,060} = 0.069$ rounded to three decimal places;

      b. $\dfrac{\binom{10}{2}\binom{8}{2}}{3,060} = \dfrac{45 \cdot 28}{3,060} = 0.412$ rounded to three decimal places.

R.93 $f(2) = \binom{5}{2}(0.70)^2(1-0.70)^{5-2} \approx 0.132$

R.95   **a.** Since $8 < 0.05(40 + 160) = 10$, the condition is satisfied.

      b. Since $10 > 0.05(100 + 60) = 8$, the condition is not satisfied.

      c. Since $12 > 0.05(68 + 82) = 7.5$, the condition is not satisfied.

R.97 $\dfrac{1}{\binom{45}{6}} = \dfrac{720}{45 \cdot 44 \cdot 43 \cdot 42 \cdot 41 \cdot 40} = \dfrac{1}{8,145,060}$

R.99   **a.** $0(0.017) + 1(0.090) + 2(0.209) + 3(0.279) + 4(0.232) + 5(0.124) + 6(0.041) + 7(0.008) + 8(0.001) = 3.203$ rounded to three decimals.

      b. $np = 8(0.40) = 3.20$

R.101   **a.** $np = 180 \cdot \dfrac{1}{9} = 20 > 10$, conditions not satisfied

       b. $n = 480 > 100$ and $np = 480 \cdot \dfrac{1}{60} = 8 < 10$, conditions are satisfied

       c. $n = 575 > 100$ and $np = 575 \cdot \dfrac{1}{100} = 5.75 < 10$, conditions are satisfied

R.103   Since 0.4713 corresponds to $z = 1.90$, $\dfrac{82.6 - \mu}{4} = 190$ and $\mu = 75$.

      Since $\dfrac{80 - 75}{4} = 1.25$, and $\dfrac{70 - 75}{4} = -1.25$, the probability is $2(0.3944) = 0.7888$.

R.105 $\dfrac{9!}{4!4!1!}(0.30)^4(0.60)^4(0.10) = 0.066$ rounded to three decimal places.

R.107   **a.** $\dfrac{\binom{5}{3}\binom{4}{0}}{\binom{9}{3}} = \dfrac{5}{42}$

b. $\dfrac{\dbinom{5}{1}\dbinom{4}{2}}{\dbinom{9}{3}} = \dfrac{5}{14}$

R.109 Use formula for the area of a triangle.

    a. $\dfrac{1}{2} \cdot 3 \cdot \dfrac{2}{3} = 1$

    b. $\dfrac{1}{2} \cdot \dfrac{3}{2} \cdot \dfrac{1}{3} = \dfrac{1}{4}$

R.111 $z = \dfrac{20 - 24.55}{3.16} = -1.44$.

    Area corresponding to $z = -1.44 = 0.4251$

    $z = \dfrac{30 - 24.55}{3.16} = 1.72$

    Area corresponding to $z = 1.72 = 0.4573$

    Probability is $0.4251 + 0.4573 = 0.8824$

R.113   **a.** Ratio is $\dfrac{\dbinom{30}{1}\dbinom{270}{11}}{\dbinom{300}{12}}$ to $\dfrac{\dbinom{30}{0}\dbinom{270}{12}}{\dbinom{300}{12}}$ or 360 to 259.

    b. $\dbinom{12}{1}(0.1)^1(0.9)^{11}$ to $\dbinom{12}{0}(0.1)^0(0.9)^{12}$ or 4 to 3.

R.115   **a.** $0.2019 + 0.3230 + 0.2584 = 0.7833$

    b. $0.1378 + 0.0551 + 0.0176 = 0.2105$

    c. $0.0176 + 0.0047 + 0.0011 + 0.0002 + 0.0000 = 0.0236$.

R.117 $\mu = 0(0.2019) + 1(0.3230) + 2(0.2584) + 3(0.1378) + 4(0.0551) + 5(0.0176) + 6(0.0047) + 7(0.0011)$
        $+ 8(0.0002) + 9(0.0000) = 1.627$.

R.119   **a.** $\sqrt{\dfrac{120 - 30}{120 - 1}} = 0.8697$

    b. $\sqrt{\dfrac{450 - 50}{400 - 1}} = 0.9366$

R.121   **a.** $\dfrac{2.3^0 e^{-2.3}}{0!} = e^{-2.3} = 0.1003$

b. $\dfrac{2.3^1 e^{-2.3}}{1!} = 0.2307$

R.123   **a.** $\dbinom{40}{2}\dbinom{20}{2}\dbinom{10}{2}\dbinom{10}{2} = 780 \cdot 190 \cdot 45 \cdot 45$

$$= 300,105,000$$

      b. $\dbinom{40}{4}\dbinom{20}{2}\dbinom{10}{1}\dbinom{10}{1} = 91,390 \cdot 190 \cdot 10 \cdot 10$

$$= 1,736,410,000$$

R.125   **a.** $0.2171 + 0.1585 + 0.0844 = 0.4600$

      b. $0.0319 + 0.0082 + 0.0013 + 0.0001 = 0.0415$

R.127   **a.** Since $np = 55 \cdot \dfrac{1}{5} = 11 > 5$ and $n(1-p) = 55 \cdot \dfrac{4}{5} = 44$ are both greater than 5, the conditions are satisfied;

      b. since $np = 105 \cdot \dfrac{1}{35} = 3$ is less than 5, the conditions are not satisfied;

      c. since $np = 210 \cdot \dfrac{1}{30} = 7$ and $n(1-p) = 210 \cdot \dfrac{29}{30} = 203$ are both greater than 5, the conditions are satisfied;

      d. since $n(1-p) = 40(0.05) = 2$ is less than 5, the conditions are not satisfied.

R.129 $\dfrac{\dbinom{7}{1}\dbinom{3}{1}\dbinom{2}{1}}{\dbinom{12}{3}} = \dfrac{7 \cdot 3 \cdot 2}{220} = \dfrac{21}{110}.$

CHAPTER 11

Problems of Estimation

11.1 Maximum error is $1.96 \cdot \dfrac{135}{\sqrt{40}} \approx 41.84$.

11.3 Maximum error is $2.575 \cdot \dfrac{3.2}{\sqrt{40}} \approx 1.30$ mm rounded to two decimal places.

11.5 Maximum error is $2.33 \cdot \dfrac{269}{\sqrt{35}} = \$106$ rounded to the nearest dollar.

11.7 $z = \dfrac{24 - 23.5}{\frac{3.3}{8}} = 1.21$ and the probability is $2(0.3869) = 0.77$ rounded to two decimal places.

11.9 $n = \left(\dfrac{2.575 \cdot 138}{40}\right)^2 = 78.92$ and $n = 79$ rounded up to the nearest integer.

11.11 $n = \left(\dfrac{2.575 \cdot 0.77}{0.25}\right)^2 = 62.90$ and $n = 63$ rounded up to the nearest integer.

11.13 **a.** $30(0.90) = 27$

b. 26; this is within one of what we could have expected.

11.15 **a.** $2.34 \pm 2.306 \cdot \dfrac{0.48}{3}$,
$2.34 - 0.37 = 1.97 < \mu < 2.34 + 0.37 = 2.71$ micrograms.

b. Maximum error is
$E = 3.355 \cdot \dfrac{0.48}{3} = 0.54$ microgram.

11.17 Maximum error is $E = 2.306 \cdot \dfrac{1,527}{3} = 1,174$ rounded to the nearest pound.

11.19 **a.** 1.771

b. 2.101

c. 2.508

d. 2.947

11.21 **a.** It is reasonable to treat the data as a sample from a normal population.

b. $31.693 \pm 2.977 \cdot \dfrac{2.156}{\sqrt{15}}$, which yields $30.04 < \mu < 33.35$ with the confidence limits rounded to two decimal places.

11.23 $12 \pm 2.821 \cdot \dfrac{2.75}{\sqrt{10}}$, which yields $9.55 < \mu < 14.45$ with the confidence limits rounded to two decimal places.

11.25 $0.15 \pm 2.447 \cdot \dfrac{0.03}{\sqrt{7}}$ which yields $0.122 < \mu < 0.178$ with the confidence limits rounded to three decimal places.

11.27 $E = 3.106 \cdot \dfrac{1.859}{\sqrt{12}} = 1.67$ fillings (rounded to two decimal places).

11.29 $n = 12$ and $s = 2.75$.
$\dfrac{11(2.75)^2}{26.757} < \sigma^2 < \dfrac{11(2.75)^2}{2.603}$ and $1.76 < \sigma < 5.65$ rounded to two decimal places.

11.31 $n = 5$ and $s = 0.381$.
$\dfrac{4(0.381)^2}{11.143} < \sigma^2 < \dfrac{4(0.381)^2}{0.484}$ and $0.052 < \sigma^2 < 1.200$ rounded to three decimal places.

11.33 $n = 12$ and $s = 1.859$
$\dfrac{(11)(1.859)^2}{26.757} < \sigma^2 < \dfrac{(11)(1.859)^2}{2.603}$

47

$1.19 < \sigma < 3.82$ rounded to two decimal places.

11.35 **a.** $\dfrac{81.0 - 70.2}{3.26} = 3.31$ rounded to two decimal places.
This is not too close to 2.75.

b. $\dfrac{14.34 - 14.26}{2.06} = 0.039$ rounded to three decimal places. This is quite close to $s = 0.0365$.

11.37 **a.** $400p > 5$ and $400(1 - p) > 5$ yields $0.0125 < p < 0.9875$.

b. $500p > 5$ and $500(1 - p) > 5$ yields $0.01 < p < 0.99$.

11.39 $\dfrac{x}{n} = 0.570$, so that

$0.570 \pm 1.96 \sqrt{\dfrac{(0.570)(0.430)}{400}}$ and $0.570 \pm$
0.0485 and $0.52 < p < 0.62$ rounded to two decimal places.

11.41 $\dfrac{x}{n} = \dfrac{56}{400} = 0.140$, so that

$0.140 \pm 2.575 \sqrt{\dfrac{(0.140)(0.860)}{400}}$ and
$0.095 < p < 0.185$.

11.43 $\dfrac{x}{n} = \dfrac{54}{120} = 0.45$;

$E = 1.645 \sqrt{\dfrac{(0.45)(0.55)}{120}} = 0.075$ rounded to three decimals.

11.45 $\dfrac{x}{n} = \dfrac{412}{1,600} = 0.2575$ and

$0.2575 \pm 1.96 \sqrt{\dfrac{(0.2575)(0.7425)}{1,600}}$. This
yields $23.6 < 100p < 27.89$ percent rounded to two decimal places.

11.47 $\dfrac{x}{n} = \dfrac{119}{140} = 0.85$ and

$0.85 \pm 2.575 \sqrt{\dfrac{(0.85)(0.15)}{140}}$. This yields 0.85
± 0.078 rounded to three decimals, and $0.772 < p < 0.928$ rounded to three decimal places.

11.49 **a.** $\dfrac{x}{n} = \dfrac{34}{100} = 0.34$ and

$0.34 \pm 1.96 \sqrt{\dfrac{(0.34)(0.66)}{100} \cdot \dfrac{(360 - 100)}{(360 - 1)}}$
This yields $0.261 < p < 0.419$ rounded to three decimal places.

b. Continuing with Exercise 11.47, we get
$0.85 \pm 0.078 \sqrt{\dfrac{350 - 140}{350 - 1}}$. This yields
0.85 ± 0.061 and $0.789 < p < 0.911$.

11.51 **a.** $\dfrac{1}{4} \left(\dfrac{1.645}{0.05} \right)^2 = 271$ rounded up to the nearest integer.

b. $n = \dfrac{1}{4} \left(\dfrac{1.96}{0.05} \right)^2 = 385$ rounded up to the nearest integer.

c. $n = \dfrac{1}{4} \left(\dfrac{2.575}{0.05} \right)^2 = 664$ rounded up to the nearest integer.

12.1   **a.** $\mu < \mu_0$ and buy the new van only if the null hypothesis can be rejected.

     b. $\mu > \mu_0$ and buy the new van unless the null hypothesis can be rejected.

12.3   Since $\bar{x} = 0.365$ second falls between 0.36 and 0.40, the psychologist will accept the null hypothesis $\mu = 0.38$ second.

     a. Since the null hypothesis is true and accepted, the psychologist will not be making an error.

     b. Since the null hypothesis is false but accepted, the psychologist will be making a Type II error.

12.5   If it erroneously rejects the null hypothesis, the testing service will be committing a Type I error; if it erroneously accepts the null hypothesis, it will be committing a Type II error.

12.7   Use the null hypothesis that the antipollution device is not effective.

12.9   **a.** $z = \dfrac{0.405 - 0.380}{\dfrac{0.08}{\sqrt{40}}} = \dfrac{0.025}{0.0126} = 1.98$

     The area corresponding to $z = 1.98$ is 0.4761.

     $z = \dfrac{0.355 - 0.380}{0.0126} = -1.98$

     and the probability of a type I error is $2(0.5000 - 0.4761) = 2(0.0239) = 0.0478$ or 0.05 rounded to two decimal places.

     b. $z = \dfrac{0.405 - 0.41}{0.0126} = -0.40$. The area corresponding to $z = -0.40$ is 0.1554.

     $z = \dfrac{0.355 - 0.41}{0.0126} = -4.37$. The area corresponding to $z = -4.37$ is 0.49997 and

$0.5000 - 0.49997$ is negligible. The probability of a Type II error is $0.5000 - 0.1554 = 0.3446$ or 0.34 rounded to two decimal places. The Type II error is increased from 0.21. See Figure 12.2.

12.11   Since we are not dealing with sample data, there is no question here of statistical inference.

12.13   To reject the null hypothesis that there is no such thing as extra sensory perception, at least 2.8 persons would have to get high scores.

12.15   The null hypothesis is $\mu = 2.6$. As it is of concern that there may be more absences than that, the alternative hypothesis should be $\mu > 2.6$.

12.17   The null hypothesis $\mu = 20$ and the alternative hypothesis $\mu > 20$, and accept the manufacturer's claim only if the null hypothesis can be rejected.

12.19   **a.** 0.05

     b. $1 - (0.95)^2 = 0.0975$

     c. $1 - (0.95)^{32} = 0.8063$

12.21   **1.** $H_0 : \mu = 12.3$ and $H_A : \mu \neq 12.3$

     2. $\alpha = 0.05$

     3. Reject H_0 if $z \leq -1.96$ or $z \geq 1.96$.

     4. $z = \dfrac{11.5 - 12.3}{\dfrac{3.8}{\sqrt{35}}} = 1.25$

     5. Null hypothesis cannot be rejected.

12.23   **1.** $H_0: \mu = 3.52$ and $H_A: \mu \neq 3.52$

     2. $\alpha = 0.05$

3. Reject H_0 if $z \leq -1.96$ or $z \geq 1.96$.

4. $z = \dfrac{3.55 - 3.52}{\frac{0.07}{\sqrt{32}}} \approx 2.42$

5. Null hypothesis must be rejected since $2.42 > 1.96$.

12.25   **1.** $H_0 \colon \mu = 83.2$ and $H_A \colon \mu > 83.2$

      2. $\alpha = 0.01$

      3. Reject H_0 if $z \geq 2.33$.

      4. $z = \dfrac{86.7 - 83.2}{\frac{8.6}{\sqrt{45}}} \approx 2.73$

      5. Null hypothesis must be rejected.

12.27   **1.** $H_0 \colon \mu = 80$ and $H_A \colon \mu < 80$

      2. $\alpha = 0.05$

      3. Reject H_0 if $t \leq -1.796$.

      4. $t = \dfrac{78.2 - 80}{\frac{7.9}{\sqrt{12}}} \approx -0.79$

      5. The null hypothesis cannot be rejected.

12.29   **1.** $H_0 \colon \mu = 0.125$ and $H_A \colon \mu > 0.125$

      2. $\alpha = 0.01$

      3. Reject H_0 if $t \geq 3.747$

      4. $t = \dfrac{13.1 - 12.5}{\frac{0.51}{\sqrt{5}}} = 2.63$

      5. The null hypotheses cannot be rejected.

12.31   **1.** $H_0 \colon \mu = 14$ and $H_A \colon \mu > 14$

      2. $\alpha = 0.05$

      3. Reject H_0 if $t \geq 1.753$

      4. $t = \dfrac{15.25 - 14}{\frac{2.70}{\sqrt{16}}} = 1.85$

      5. The null hypothesis must be rejected.

12.33   Normal probability plot indicates that population is not normal.

12.37 $s_p = 3.084$ and $t = 2.29$. The null hypothesis must be rejected.

12.39 $s_p = 19.10$ and $t = -2.12$. The null hypothesis cannot be rejected.

12.41   The *p*-value is 0.0744. Since this is less than 0.10, the null hypothesis could have been rejected.

12.43   The *p*-value is 0.2446 and this is the lowest level of significance at which the null hypothesis could have been rejected.

12.45   **1.** $H_0 \colon \mu_1 - \mu_2 = 0$

      $H_A \colon \mu_1 - \mu_2 \neq 0$

      2. $\alpha = 0.05$

      3. Reject the null hypothesis if $z \leq -1.96$ or $z \geq 1.96$, where

$$z = \dfrac{\bar{x}_1 - \bar{x}_2}{\sqrt{\dfrac{s_1^2}{n_1} + \dfrac{s_2^2}{n_2}}}$$

      4. Substituting the given values of n_1, n_2, $\bar{x}_1, \bar{x}_2, s_1$ and s_2 into the formula for z, we get

$$z = \dfrac{64.20 - 71.41}{\sqrt{\dfrac{(16.00)^2}{80} + \dfrac{(22.13)^2}{100}}} = -2.53$$

      5. Since $z = -2.53$ is less than -1.96, it follows that the null hypothesis must be rejected. That is, the difference between the two sample means is too large to be attributed to chance, and we conclude that there is a real difference between the two population means.

12.47   **1.** $H_0 \colon \mu_1 - \mu_2 = -0.05$.

      $H_A \colon \mu_1 - \mu_2 < -0.05$.

      2. $\alpha = 0.05$

50

3. Reject the null hypothesis if $t \le -1.645$, where

$$t = \frac{(x_1 - x_2) - \delta}{\sqrt{\frac{(n_1 - 1)s_1^2 + (n_2 - 1)s_2^2}{n_1 + n_2 - 2} \cdot (\frac{1}{n_1} + \frac{1}{n_2})}}$$

4. Substituting the $\bar{x}_1 = 0.083$, $n_1 = 25$, $s_1 = 0.003$, $\bar{x}_2 = 0.136$, $n_2 = 25$, and $s_2 = 0.002$ into the formula for t, we get

$$t = \frac{(0.083 - 0.136) - (-0.05)}{\sqrt{\frac{24(0.003)^2 + 24(0.002)^2}{25 + 25 - 2} (\frac{1}{25} + \frac{1}{25})}} = -4.16$$

5. Since -4.16 is less than -1.645 the null hypothesis must be rejected. The difference is significant. The claim has been substantiated.

13.1　　**1.** H_0: $\sigma = 0.0100$ and H_A: $\sigma < 0.0100$

　　　　2. $\alpha = 0.05$

　　　　3. Reject H_0 if $\chi^2 \le 3.325$

　　　　4. $\chi^2 = \dfrac{9(0.0086)^2}{(0.010)^2} = 6.66$

　　　　5. Null hypothesis cannot be rejected.

13.3　　The p-value is $2(0.0092) = 0.0184$. Since $0.0184 \le 0.03$, the null hypothesis must be rejected.

13.5　　**1.** H_0: $\sigma = 0.80$ and H_A: $\sigma < 0.80$

　　　　2. $\sigma = 0.01$

　　　　3. Reject H_0 if $z \le -2.33$.

　　　　4. $z = \dfrac{0.74 - 0.80}{\frac{0.80}{\sqrt{80}}} = -0.674$

　　　　5. Null hypothesis cannot be rejected.

13.7　　**1.** H_0: $\sigma_1 = \sigma_2$ and H_A: $\sigma_1 < \sigma_2$

　　　　2. $\alpha = 0.05$

　　　　3. Reject H_0 if $F \ge 2.72$.

　　　　4. $F = \dfrac{(4.4)}{(2.6)^2} = 2.86$

　　　　5. The null hypothesis must be rejected.

14.1 **1.** $H_0: p = 0.05$ and $H_A: p > 0.05$

2. $\alpha = 0.01$

3. The test statistic is the observed number of couples who take such a cruise within a year's time.

4. $x = 3$ and the probability of 3 or more successes is 0.043.

5. Since 0.043 is less than 0.01, the null hypothesis cannot be rejected.

14.3 **1.** $H_0: p = 0.50$ and $H_A: p \neq 0.50$

2. $\alpha = 0.10$

3. Test statistic is x, the number of persons in the sample who are opposed to capital punishment.

4. From Table V, with $n = 20$, $x = 14$, $p = 0.5$. The p-value of this two-tailed test is $2(0.058) = 0.116$.

5. The null hypothesis cannot be rejected.

14.5 a. **1.** $H_0: p = 0.36$ and $H_A: p < 0.36$
2. $\alpha = 0.05$
3. Reject H_0 if $z \leq -1.645$.
4. $z = \dfrac{94 - 300(0.36)}{\sqrt{300(0.36)(0.64)}} \approx -1.68$
5. Null hypothesis must be rejected.

b. Using continuity correction, steps 1, 2, and 3 are the same.
4. $z = \dfrac{94.5 - 300(0.36)}{\sqrt{300(0.36)(0.64)}} \approx -1.62$
5. Null hypothesis cannot be rejected.

14.7 **1.** $H_0: p = 0.95$ and $H_A: p \neq 0.95$

2. $\alpha = 0.01$

3. Reject H_0 if $z \leq -2.575$ or $z = \geq 2.575$

4. $z = \dfrac{464.5 - 500(0.95)}{\sqrt{500(0.95)(0.05)}} = -2.15$

5. The null hypothesis cannot be rejected.

14.9 **1.** $H_0: p_1 = p_2$ and $H_A: p_1 > p_2$

2. $\alpha = 0.05$

3. Reject H_0 if $z \geq 1.645$

4. $\hat{p} = \dfrac{54 + 33}{250} = 0.348$ and

$z = \dfrac{0.36 - 0.33}{\sqrt{(0.348)(0.652)\left(\frac{1}{150} + \frac{1}{100}\right)}} = 0.49$

5. Null hypothesis cannot be rejected.

14.11 **1.** $H_0: p_1 = p_2$ and $H_A: p \neq p$

2. $\alpha = 0.05$

3. Reject H_0 if $z \leq -1.96$ or $z \geq 1.96$.

4. $z = \dfrac{0.22 - 0.275}{\sqrt{(0.25)(0.75)\left(\frac{1}{100} + \frac{1}{120}\right)}} = -0.94$

5. Null hypothesis cannot be rejected.

14.13 **1.** $H_0: p_1 = p_2$ and $H_A: p \neq p$

2. $\alpha = 0.05$

3. Reject H_0 if $z \leq -1.96$ or $z \geq 1.96$.

4. $\dfrac{x_1}{n_1} = \dfrac{62}{100} = 0.62$, $\dfrac{x_2}{n_2} = \dfrac{44}{100} = 0.44$, and

$\hat{p} = \dfrac{62 + 44}{100 + 100} = 0.53$

$z = \dfrac{0.62 - 0.44}{\sqrt{(0.53)(0.47)\left(\frac{1}{100} + \frac{1}{100}\right)}} \approx 2.55$

5. The null hypothesis must be rejected. The difference between the sample proportions is not significant.

14.15 $\chi = \dfrac{67}{46.4} + \dfrac{64}{63.6} + \cdots + \dfrac{37}{20.7} - 400$
       $= 40.89$

14.17 H_0: The probabilities about the three
       response categories (part-time
       employment, full-time employment,
       no
       employment) are all equal regardless
       of the number of children.

   H_A: The probabilities for at least one of
       the response categories are not all the
       same.

14.21   **1.** H_0: The probabilities for the three
       response categories are the same
       for
       all four ranks.
       H_A: The probabilities for at least one
       response category are not the same
       for all four ranks.

   2. $\alpha = 0.01$

   3. Reject H_0 if $\chi^2 \geq 16.812$.

   4. The expected frequencies are 10.8, 18.9,
       13.5, 10.8 for the first row, 24.2, 42.35,
       30.25, and 24.2 for the second row, and
       45, 78.75, 56.25, and 45 for the third
       row.
       $\chi^2 = \dfrac{(8-10.8)^2}{10.8} + \cdots + \dfrac{(52-45)^2}{45}$
       $= 20.72$

   5. The null hypothesis must be rejected.

14.23   **1.** H_0: Students' interest and ability in
       studying a foreign language are
       independent.
       H_A: The two variables are not
       independent.

   2. $\alpha = 0.05$

   3. Reject H_0 if $\chi^2 \geq 9.488$

   4. The expected values for the first row are
       16.364, 23.182, and 20.455, for the
       second row are 21.818, 30.909, and
       27.273 and for the third row are 21.818,
       30.909, and 27.273 for the third row.
       $\chi^2 = 26.77$

   5. The null hypothesis must be rejected.

14.27   The new first column becomes 28, 74, and
       18.

   1. H_0: The handicaps do not affect the
       performance.
       H_A: The handicaps do affect the
       performance.

   2. $\alpha = 0.05$

   3. Reject H_0 if $\chi^2 \geq 5.991$, which is the
       value of $\chi^2_{0.05}$ for 2 degrees of freedom.

   4. The expected values for the first row are
       24 and 40, those for the second row are
       78 and 130, and those for the third row
       are 18 and 30.
       $\chi^2 = 1.39$

   5. The null hypothesis cannot be rejected.

14.29   **1.** H_0: There is no relationship between the
       fidelity and the selectivity of the
       radios.
       H_A: There is a relationship between the
       two variables.

   2. $\alpha = 0.01$

   3. Reject the null hypothesis if
       $\chi^2 \geq 13.277$

   4. The expected frequencies are 15.00,
       22.11, and 12.89 for the first row, 33.6,
       49.52, and 28.88 for the second row, and
       8.40, 12.38, and 7.22 for the third row.
       $\chi^2 = 52.72$

   5. The null hypothesis must be rejected.

14.31 $\chi^2 = 1.23$ and $z = -1.11$ from Exercise 14.12

   $z^2 = 1.23$ which equals χ^2.

14.33   **1.** H_0: The probability of a response
       favoring the candidate is the same for all
       five unions.
       H_A: The probabilities are not all the
       same.

   2. $\alpha = 0.01$

3. Reject H_0 if $\chi^2 \geq 13.277$.

4. The expected frequencies for the first row are all 78, and those for the second row are all 22. $\chi^2 = 16.55$.

5. The null hypothesis must be rejected.

14.35   Let r_i be the total of the observed frequencies for the ith row, c_j the total of the observed frequencies for the jth column, e_{ij} the expected frequency for the ith row and the jth column, and n the grand total for the entire table.

$$\sum_i e_{ij} = \sum_j \frac{r_i \cdot c_j}{n}$$
$$= \frac{c_j}{n} \cdot \sum r_i$$
$$= \frac{c_j}{n} \cdot n$$
$$= c_j$$

14.37   1. H_0: Data constitute sample from a binomial population with $n = 4$ and $p = 0.50$.
       H_A: Data do not constitute sample from a binomial population with $n = 4$ and $p = 0.50$.

2. $\alpha = 0.01$

3. Reject H_0 if $\chi^2 \geq 13.277$

4. The expected frequencies are 10, 40, 60, 40, and 10. $\chi^2 = 2.37$

5. The null hypothesis cannot be rejected.

14.39 $\mu = 0 \cdot \dfrac{2}{300} + 1 \cdot \dfrac{10}{300} + \ldots + 4 \cdot \dfrac{119}{300} = 3.2 = 4p,$
   so that $p = \dfrac{3.2}{4} = 0.8$.

1. H_0: Data constitute sample from a binomial population with $n = 4$ and $p = 0.8$.
       H_A: Data do not constitute sample from binomial population with $n = 4$ and $p = 0.8$.

2. $\alpha = 0.05$

3. Reject H_0 if $\chi^2 \geq 5.991$

4. The probabilities are 0.002, 0.026, 0.154, 0.410, and 0.410, so that the expected frequencies for 0, 1, 2, 3, and 4 are 0.6, 7.8, 46.2, 123, and 123. Combine 0 and 1.
$$\chi^2 = 6.82$$

5. The null hypothesis must be rejected.

Review Exercises for Chapters 11, 12, 13, and 14

R.131 $n = (0.22)(0.78)\left(\dfrac{1.96}{0.035}\right)^2 = 539$ rounded up to the nearest integer.

R.135 **1.** H_0: $\mu = 78$ and H_A: $\mu > 78$.

    2. $\alpha = 0.01$

    3. Reject H_0 if $t \geq 3.747$.

    4. $t = \dfrac{82.2 - 78}{1.351\sqrt{5}} = 6.95$

    5. The null hypothesis must be rejected.

R.137 **1.** H_0: The probabilities of the responses are the same regardless of the number of children.
H_A: The probabilities of the responses are not all the same at least for one of the numbers of children.

    2. $\alpha = 0.05$

    3. Reject the null hypothesis if $\chi^2 \geq 9.488$.

    4. The expected frequencies for the first row are 44.4, 38.6, and 16.9, those for the second row are 60.9, 52.9, and 23.2, and those for the third row are 54.7, 47.5, and 20.8.
$\chi^2 = \dfrac{(48 - 44.4)^2}{44.4} + \cdots + \dfrac{(20 - 20.8)^2}{20.8}$
$= 3.97$

    5. The null hypothesis cannot be rejected.

R.139 First two steps same as before.

    3. Reject H_0 if $\chi^2 \geq 6.635$.

    4. The expected frequencies for the first row are 67.8 and 45.2, those for the second row are 232.2 and 154.8.
$\chi^2 = \dfrac{(81 - 67.8)^2}{67.8} + \cdots + \dfrac{(168 - 154.8)^2}{154.8}$
$= 8.30$
rounded to two decimals.

    5. The null hypothesis must be rejected.

R.141 $\mu = 1.4 = 7p$, so that $p = 0.20$. Combine 3, 4, and 5.

    1. H_0: $p = 0.20$ and H_A: $p \neq 0.20$.

    2. $\alpha = 0.05$

    3. Reject H_0 if $\chi^2 \geq 5.991$.

    4. $\chi^2 = \dfrac{(12 - 10.5)^2}{10.5} + \cdots + \dfrac{(9 - 7.4)^2}{7.4}$
$= 0.91$

    5. The null hypothesis cannot be rejected.

R.143 **1.** H_0: The three programs are equally effective.
H_A: The three programs are not equally effective.

    2. $\alpha = 0.05$

    3. Reject H_0 if $\chi^2 \geq 2.920$.

    4. $\chi = \dfrac{(86 - 82.9)^2}{82.9} + \cdots + \dfrac{(38 - 33.1)^2}{33.1} = 1.659$

    5. The null hypothesis cannot be rejected.

R.145 The normal probability plot reveals a linear pattern.

R.147 Since we are not dealing with samples, there is no question here of statistical significance.

R.149 $\dfrac{10.4}{1 + \frac{2.575}{\sqrt{120}}} < \sigma < \dfrac{10.4}{1 - \frac{2.575}{\sqrt{120}}}$ yields $8.42 < \sigma < 13.59$.

R.151 $\hat{p}_1 = \dfrac{23}{80} = 0.2875$, $\hat{p}_2 = \dfrac{19}{80} = 0.2375$, and
$\hat{p} = \dfrac{19 + 23}{80 + 80} = 0.2625$

    1. H_0: $p_1 = p_2$ and H_A: $p_1 \neq p_2$

    2. $\alpha = 0.05$

    3. Reject H_0 if $z \leq -1.96$ or $z \geq 1.96$.

4. $z = \dfrac{0.2875 - 0.2375}{\sqrt{(0.2625)(0.7375)\left(\frac{1}{80} + \frac{1}{80}\right)}}$

$= 0.72$

rounded to two decimals

5. The null hypothesis cannot be rejected.

R.153   **1.** $H_0: p_1 = p_2 = p_3$ and $H_A: p_1, p_2,$ and p_3 are not all equal.

     2. $\alpha = 0.01$

     3. Reject H_0 if $\chi^2 \geq 9.210$.

     4. The expected frequencies for the first row are all 72; those for the second row are all 28.

$\chi^2 = \dfrac{(63 - 72)^2}{72} + \cdots + \dfrac{(31 - 28)^2}{28} = 11.61$

     5. The null hypothesis must be rejected.

R.155 $n = \left[\dfrac{2.575(3.4)}{1.2}\right]^2 = 54$ rounded up to the nearest integer

R.157   **a.** 0.0375, 0.1071, 0.2223, 0.2811, 0.2163, 0.1013, and 0.0344

     b. 3, 8.57, 17.78, 22.49, 17.30, 8.10, and 2.75

     c.   **1.** H_0: population sampled is normal and H_A: population sampled is not normal.

        2. $\alpha = 0.05$

        3. Combining first two classes and last two classes, reject H_0 if $\chi^2 \geq 5.991$,

        4. $\chi^2 = \dfrac{(13 - 11.57)^2}{11.57} + \cdots + \dfrac{(11 - 10.85)^2}{10.85}$

$= 1.27$

        5. The null hypothesis cannot be rejected.

R.159 $n_1 = 10,\ n_2 = 8,\ s_1 = 4.395,$ and $s_2 = 1.637$.

     1. $H_0: \sigma_1 = \sigma_2$ and $H_A: \sigma_1 \neq \sigma_2$.

     2. $\alpha = 0.02$

3. Reject H_0 if $F \geq 6.72$.

4. $F = \dfrac{(4.395)^2}{(1.637)^2} = 7.21$

5. The null hypothesis must be rejected.

R.161   **a.** $s = 5.10$

     b. $\dfrac{14}{2.85} = 4.91$

R.163   **1.** $H_0: \sigma_1 = \sigma_2$ and $H_A: \sigma_1 \neq \sigma_2$

     2. $\alpha = 0.10$

     3. Reject H_0 if $F \geq 5.05$.

     4. $F = \dfrac{(3.3)^2}{(2.1)^2} = 2.47$

     5. The null hypothesis cannot be rejected.

R.165   **1.** $H_0: \sigma = 1.0$ and $H_A: \sigma > 1.0$

     2. $\alpha = 0.01$

     3. Reject H_0 if $\chi^2 \geq 21.666$, the value of $\chi^2_{0.01}$ for 9 degrees of freedom.

     4. $\chi^2 = \dfrac{9(1.28)^2}{1.0^2} = 14.75$

     5. The null hypothesis cannot be rejected.

R.167   **a.** $\mu > \mu_0$ and replace the old machines only if the null hypothesis can be rejected.

     b. $\mu < \mu_0$ and replace the old machines unless the null hypothesis can be rejected.

57

CHAPTER 15

Analysis of Variance

15.1 **a.**

$$ns_{\bar{x}}^2 = 6 \cdot \frac{(3.3 - 2.7)^2 + (2.6 - 2.7)^2 + (2.2 - 2.7)^2}{2}$$

$$= 1.86$$

$$\frac{1}{3}(s_1^2 + s_2^2 + s_3^2) = \frac{4.796 + 2.784 + 2.064}{3} = 3.215$$

$$F = \frac{1.86}{3.215} = 0.58.$$

b. 1. $H_0: \mu_1 = \mu_2 = \mu_3$ and H_A: the μ's
are not all equal.
2. $\alpha = 0.05$
3. Reject H_0 if $F \geq 3.68$.
4. $F = 0.58$
5. The null hypothesis cannot be rejected.

15.3 **a.**

$$ns_{\bar{x}}^2 = 3 \cdot \frac{(63 - 60)^2 + (58 - 60)^2 + (58 - 60)^2 + (61 - 60)^2}{3} = 18$$

$$\frac{1}{4}(s_1^2 + s_2^2 + s_3^2 + s_4^2) = \frac{1}{4}(7 + 19 + 3 + 3) = 8,$$

and $F = \frac{18}{8} = 2.25.$

b. 1. $H_0: \mu_1 = \mu_2 = \mu_3 = \mu_4$ and H_A: the
μ's are not all equal.
2. $\alpha = 0.01$
3. Reject H_0 if $F \geq 7.59$
4. $F = 2.25$
5. The null hypothesis cannot be rejected.

15.5 The scores for School 2 are much more
variable than those for the other two schools.

15.7 The three kinds of tulips should have been
assigned at random to the twelve locations in
the flower bed.

15.9 This is controversial, and statisticians argue
about the appropriateness of discarding a
proper randomization because it happens to
possess some undesirable property. For the
situation described here it is likely that the
analysis of variance will not be performed
as is.

15.15 The degrees of freedom for treatments and
error are 3 and 20, the sums of squares for
treatments and error are 32.34 and 20.03, the
mean squares for treatments and error are
10.78 and 1.00, and F equals 10.78.

1. $H_0: \mu_1 = \mu_2 = \mu_3 = \mu_4$ and H_A: the μ's
are not all equal.

2. $\alpha = 0.01$

3. Reject H_0 if $F \geq 4.94$, the value of $F_{0.01}$
for 3 and 20 degrees of freedom.

4. $F = 10.78$

5. The null hypothesis must be rejected.
The differences among the sample means
cannot be attributed to chance.

15.17 The degrees of freedom for treatments and
error are 3 and 19, the sums of squares for
treatments and error are 7,669.19 and
4,152.55, the mean squares for treatments
and error are 2,556.40 and 218.56, and the
value of F is 11.70.

1. $H_0: \mu_1 = \mu_2 = \mu_3 = \mu_4$ and H_A: the μ's
are not all equal.

2. $\alpha = 0.01$

3. Reject H_0 if $F \geq 5.01$

4. $F = 11.70$

5. The null hypothesis must be rejected.
The differences among the sample means
cannot be attributed to chance.

15.19 The degrees of freedom for treatments and
error are 2 and 12, the sums of squares for
treatments and error are 79.65 and 149.95,
the mean squares for treatments and error are
39.82 and 12.50, and the value of F is 3.19.

1. $H_0: \mu_1 = \mu_2 = \mu_3$ and H_A: the μ's are not all equal.

2. $\alpha = 0.05$

3. Reject H_0 if $F \geq 3.89$, the value of $F_{0.05}$ for 2 and 12 degrees of freedom.

4. $F = 3.19$

5. The null hypothesis cannot be rejected. The differences among the three sample means are not significant.

15.21 The degrees of freedom for treatments and error are 7 and 40, the sums of squares for treatments and error are 12,696.20 and 7,818.70, the mean squares for treatments and error are 1,813.74 and 195.47, and the value of F is 9.28.

1. $H_0: \mu_1 = \mu_2 = \cdots = \mu_8$ and H_A: the μ's are not all equal.

2. $\alpha = 0.05$

3. Reject H_0 if $F \geq 2.25$

4. $F = 9.28$

5. The null hypothesis must be rejected.

15.23 The difference between the mean of two treatments is significant if $| \bar{x}_i - \bar{x}_j | \frac{q_\alpha}{\sqrt{n}} \cdot s$

where $q_\alpha = 4.64$, $n = 6$, $s = 1.0$ from Exercise 15.15. $| \bar{x} - \bar{x}_j | \geq 1.89$

Treatment	32	28	24	20
$\bar{x}$	19.83	21.73	22.25	22.97

32 − 28	$|19.83 - 21.73| = 1.90$	significant
32 − 24	$|19.83 - 22.25| = 2.42$	significant
32 − 20	$|19.83 - 22.97| = 3.14$	significant
28 − 24	$|21.73 - 22.97| = 1.24$	not significant
28 − 20	$|21.73 - 22.97| = 1.24$	not significant
24 − 20	$|22.25 - 22.97| = 0.72$	not significant

32   28   24   20

▬▬▬▬▬▬▬

The three underlined treatments are not significant.

15.25 There are six possible pairs of means. The three underlined pairs are not significant. Only the differences between Mr. Brown and Mr. Black, Mr. Brown and Mrs. Smith, and Ms. Jones and Mrs. Smith are significant.

15.29 She might consider only programs of the same length, or she might use the program lengths as blocks and perform a two-way analysis of variance.

15.31 The degrees of freedom for treatments (diet foods), blocks (laboratories), and error are 2, 3, and 6. The sums of squares for treatments, blocks, and error are 0.49, 0.54, and 0.22, the mean squares for treatments, blocks and error are 0.245, 0.18, and 0.037, and the values of F for treatments and blocks are 6.62 and 4.86.

1. $H_0: \alpha_1 = \alpha_2 = \alpha_3 = 0;$
$\beta_1 = \beta_2 = \beta_3 = \beta_4 = 0$
H_A: The α's are not all equal to zero and the β's are not all equal to zero.

2. $\alpha = 0.01$ for both tests.

3. Reject H_0 for treatments if $F \geq 10.9$ and for blocks of $F \geq 9.78$.

4. $F = 6.62$ for treatments and $F = 4.86$ for blocks.

5. Neither null hypothesis can be rejected.

15.33 The degrees of freedom for treatments (threads), blocks (measuring instruments), and error are 4, 3, and 12. The sums of squares for treatments, blocks, and error are 70.18, 3.69, and 25.31. The mean squares for treatments, blocks, and error are 17.54, 1.23, and 2.11. The values of F for treatments and blocks are 8.31 and 0.58.

1. H_0: The α's are all equal to zero and the B's are equal to zero.
H_A: The α's are not all equal to zero and the β's are all equal to zero.

2. $\alpha = 0.05$ for both tests.

3. Reject H_0 for treatments if $F \geq 3.26$ and for blocks if $F \geq 3.49$.

4. $F = 8.31$ for treatments and $F = 0.58$ for blocks.

5. The null hypothesis for treatments must be rejected; the null hypothesis for blocks cannot be rejected.

15.35 The degrees of freedom for water temperature, detergents, interaction, and error are 2, 2, 4, and 18. The sums of squares for water temperature, detergents, interaction, and error are 34.3, 442.7, 146.6, and 246.7. The mean squares for water temperature, detergents, interaction, and error are 17.1, 221.4, 36.6, and 13.7. The values of F for water temperature, detergents, and interaction are 1.25, 16.16, and 2.67.

1. H_0: The water temperature effects are all equal to zero. The detergent effects are all equal to zero. The interaction effects are all equal to zero.

H_A: The water temperature effects are not all equal to zero. The detergent effects are not all equal to zero. The interaction effects are not all equal to zero.

2. $\alpha = 0.01$ for each test.

3. Reject H_0 for water temperatures, if $F \geq 6.01$ for detergents if $F \geq 6.01$, and for interaction if $F \geq 4.58$.

4. $F = 1.25$ for water temperature, $F = 16.6$; for detergents, and $F = 2.67$ for interaction.

5. The null hypothesis for detergents must be rejected; the other two hypotheses cannot be rejected.

15.37 $15 \cdot 4 \cdot 3 = 180$

15.39 $A_L B_L C_L D_L$, $A_L B_L C_L D_H$, $A_L B_L C_H D_L$, $A_L B_L C_H D_H$, $A_L B_H C_L D_L$, $A_L B_H C_L D_H$, $A_L B_H C_H D_L$, $A_L B_H C_H D_H$, $A_H B_L C_L D_L$, $A_H B_L C_L D_H$, $A_H B_L C_H D_L$, $A_H B_L C_H D_H$, $A_H B_H C_L D_L$, $A_H B_H C_L D_H$, $A_H B_H C_H D_L$, and $A_H B_H C_H D_H$.

15.43 Letting I stand for Independent, R for Republican, and D for Democrat, we are given

	Teacher	Lawyer	Doctor
Easterner	I		
Southerner	R	D	
Westerner			

Completing the Latin Square, which is not difficult, leads to the result that the doctor who is a Westerner is a Republican.

15.45 Since 2 already appears with 1, 3, 4, and 6, it must appear together with 5 and 7 on Thursday. Since 4 already appears with 1, 2, 5, and 6, it must appear together with 3 and 7 on Tuesday. Since 5 already appears together with 1, 2, 4, and 7, it must appear together with 3 and 6 on Saturday.

16.1

$y = 10 - 1/2x$	$y - \hat{y}$	$(y - \hat{y})^2$
$\hat{y} = 10 - 1/2(6) = 7$	$5 - 7 = -2$	4
$\hat{y} = 10 - 1/2(12) = 4$	$6 - 4 = 2$	4
$\hat{y} = 10 - 1/2(18) = 1$	$1 - 1 = 0$	0
		$\Sigma(y - \hat{y})^2 = 8$

$y = 8 - 1/3x$	$y - \hat{y}$	$(y - \hat{y})^2$
$\hat{y} = 8 - 1/3(6) = 6$	$5 - 6 = -1$	1
$\hat{y} = 8 - 1/3(12) = 4$	$6 - 4 = 2$	4
$\hat{y} = 8 - 1/3(18) = 2$	$1 - 2 = -1$	1
		$\Sigma(y - \hat{y})^2 = 6$

$6 < 8$ so $y = 8 - 1/3x$ provides a better fit.

16.3   **a.** The points are fairly dispersed, but the overall pattern is that of a straight line.

     c. The estimate is about 12 or 13.

16.5   The two normal equations are $100 = 10a + 525b$ and $5,980 = 525a + 32,085b$. Their solution is $a = 1.526$ and $b = 0.161$.

16.7 $\hat{y} = 19.6$

16.9   **a.** $\hat{y} = 2.039 - 0.102x$

     b. $\hat{y} = 1.529$

     c. On a very hot day the chlorine would dissipate much faster.

16.11   The sum of the squares of the errors is 20.94, which is less than either 44 or 26.

16.13 $\hat{y} = 0.4911 + 0.2724x$

16.15 $\hat{y} = 10.83$

16.17 $\hat{y} = 2.66 + 0.6\mu$; $\hat{y} = \$4.46$ million.

16.19   **a.** $a = 12.447$ and $b = 0.898$

      b. $a = 0.4898$ and $b = 0.2724$

16.21   **a.** $S_{xx} = 88$, $S_{yy} = 92.83$, $S_{xy} = 79$, and $s_e = 2.34$.

      b. 1. H_0: $\beta = 1.5$ and H_A: $\beta < 1.5$
         2. $\alpha = 0.05$
         3. Reject H_0 if $t \geq -2.132$ where 2.132 is the value of $t_{0.05}$ for 4 degrees of freedom.
         4. $t = -2.41$
         5. The null hypothesis must be rejected.

16.23   **1.** H_0: $\beta = 3.5$ and H_A: $\beta > 3.5$

      2. $\alpha = 0.01$

      3. Reject H_0 if $t \geq 3.143$.

      4. $t = 2.68$

      5. The null hypothesis cannot be rejected.

16.25   **1.** H_0: $\beta = -0.15$ and H_A: $\beta \neq -0.15$

      2. $\alpha = 0.01$

      3. Reject H_0 if $t \leq -4.032$ or $t \geq 4.032$.

      4. $t = 4.17$

      5. The null hypothesis must be rejected.

16.27   **a.** $S_{yy} = 74$, and $s_e = 1.070$.

      b. 1. H_0: $\beta = 0.40$ and H_A: $\beta < 0.40$
         2. $\alpha = 0.05$
         3. Reject H_0 if $t \leq -1.812$.
         4. $t = -3.48$
         5. The null hypothesis must be rejected.

16.29 $a + bx_0 = 14.09$
      $13.16 < \mu_{y|50} < 15.02$

16.31   **a.** $7.78 < \mu_{y|60} < 14.64$

      b. $0.43 - 21.99$

16.33   **a.** $0.553 < \mu_{y \cdot 5} < 1.575$

         b. $0.152 - 1.976$

16.35   **a.** $\hat{y} = 198 + 37.2x_1 - 0.120x_2$

         b. $\hat{y} = 198 + 37.2(0.14) - 0.120(1,100)$
            $= 71.2$

16.37   **a.** $\hat{y} = -2.33 + 0.900x_1 + 1.27x_2 + 0.900x_3$

         b. $\hat{y} = -2.33 + 0.900(12.5) + 1.27(25) + 0.900(15)$
            $= 54.17\%$

16.39   The two normal equations are

         $11.9286 = 5(\log a) + 30(\log b)$

         $75.2228 = 30(\log a) + 220(\log b)$

         $\log b = 0.0913$

         $\log a = 1.8379$

         a. $\log \hat{y} = 1.8379 + 0.8913x$

         b. $\hat{y} = 68.9(1.234)^x$

         c. $\log \hat{y} = 1.8379 + 0.0913(5) = 2.2944$

            $\hat{y} = 197.0$

16.41 $\hat{y} = (101.17)(0.9575)^x$

16.43 $\hat{y} = (1.178)(2.855)^x$

16.45 $\hat{y} = (18.99)(1.152)^x$

16.47 $\hat{y} = 384.4 - 36.0x + 0.896x^2$
         $\hat{y} = 384.4 - 36.0(12) + 0.896(12)^2 = 81.4$

17.1 $r = \dfrac{23.6}{\sqrt{344(1.915)}} = 0.92$; the printout yields

$\sqrt{0.845} = 0.919$.

17.3 $(0.78)^2 \, 100 = 60.8\%$

17.5 $r = -0.01$

17.7 $r = -0.99$

$(-0.99)^2 \, 100 = 98.01\%$

17.9 No correction is needed. The correlation coefficient does not depend on the units of measurement.

17.11 **a.** Positive correlation

b. negative correlation

c. negative correlation

d. no correlation

e. positive correlation

17.13 $\left(\dfrac{0.41}{0.29}\right)^2 = 1.999$

The first relationship is just about twice as strong as the second.

17.15 Correlation does not necessarily imply causation. Actually, both variables (foreign language degrees and railroad track) depend on other variables, such as population growth and economic conditions in general.

17.17 Labeling the rows $x = -1, 0,$ and 1, and the columns $y = -1, 0,$ and 1, we get

$\sum x = -22,\ \sum y = -8,\ \sum x^2 = 78,$

$\sum y^2 = 106,\ \text{and} \sum xy = -39,$ so that

$S_{xx} = 75.45,\ S_{yy} = 105.66,\ S_{xy} = 39.93$ and

$r = -0.45.$

17.19 **a.** $z = 2.35$; the null hypothesis must be rejected.

b. $z = 1.80$; the null hypothesis cannot be rejected.

c. $z = 2.29$; the null hypothesis must be rejected.

17.21 **a.** $z = 1.22$; the null hypothesis cannot be rejected.

b. $z = 0.50$; the null hypothesis cannot be rejected.

17.23 $n = 12$ and $r = 0.77$

(a) **1.** $H_0: \rho = 0$ and $H_A: \rho \neq 0$

2. $\alpha = 0.01$

3. Reject H_0 if $t \leq -3.169$ or $t \geq 3.169$.

4. Use $t = \dfrac{r\sqrt{n-2}}{\sqrt{1-r^2}} = 3.82$

5. $3.82 \geq 3.169$. The null hypothesis must be rejected.

$n = 16$ and $r = 0.49$

(b) **1.** $H_0: \rho = 0$ and $H_A: \rho \neq 0$

2. $\alpha = 0.01$

3. Reject H_0 if $t \leq -2.977$ or $t \geq 2.977$.

4. Use $t = \dfrac{r\sqrt{n-2}}{\sqrt{1-r^2}} = 2.10$

5. $2.10 \leq 2.977$. The null hypothesis cannot be rejected.

17.25 **1.** $H_0: \rho = 0.50$ and $H_A: \rho > 0.50$

2. $\alpha = 0.05$

3. Reject H_0 if $z \geq 1.645$.

          4. $z = 0.93$

          5. The null hypothesis cannot be rejected.

17.27   **a.** $0.533 < \mu_Z < 1.665$ and $0.49 < \rho < 0.93$;

        b. $-0.637 < \mu_Z < 0.147$ and $-0.56 < \rho < 0.15$

        c. $0.365 < \mu_Z < 0.871$ and $0.35 < \rho < 0.70$.

17.29 $R^2 = 0.3320$ and $R = 0.576$.

17.31 $R^2 = 0.984$ and $R = 0.992$.

17.33 $r_{12} = 0$, $r_{13} = 0.20$, $r_{23} = -0.98$, and $r_{12.3} = -1.00$.

18.1 Since none of the values equals 9.00, none is discarded and the sample size is 15.

1. H_0: $\mu = 9.00$

   H_A: $\mu > 9.00$

2. $\alpha = 0.05$

3. The criterion may be based on the number of plus signs or the number of minus signs denoted by x. Reject the null hypothesis if the probability of getting x or more plus signs is less than or equal to 0.05.

4. Replacing each value greater than 9.00 with a plus sign and each value less than 9.00 with a minus sign, we get

   $$+ + - - + + + + + + - + + - +$$

   where there are 11 plus signs. Table V shows that for $n = 15$ and $p = 0.50$ the probability of 11 or more plus signs is equal to $0.042 + 0.014 + 0.03 = 0.059$.

5. Since 0.059 exceeds 0.05, the null hypothesis cannot be rejected.

18.3

1. H_0: $\tilde{\mu} = 110$ and H_A: $\tilde{\mu} > 110$

2. $\alpha = 0.01$

3. The test statistic, x, is the number of packages weighing more than 110 grams.

4. $x = 14$ (out of $n = 18$) and the p-value is 0.016.

5. The null hypothesis cannot be rejected.

18.5

1. H_0: $\tilde{\mu} = 278$ and H_A: $\tilde{\mu} > 278$

2. $\alpha = 0.05$

a. 3. The test statistic, x, is the number of scores greater than 278.

4. $x = 10$ and the p-value is 0.047.

5. The null hypothesis must be rejected.

b. 3. Reject H_0 if $z \geq 1.645$.

Since the two scores that equal 278 must be discared, $n = (15 - 2) = 13$; $np = 13(0.5) = 6.5$; $\sigma = \sqrt{13(0.5)(0.5)} = 1.803$

4. $z = \dfrac{10 - 6.5}{1.803} = 1.94.$

5. The null hypothesis must be rejected.

18.7

1. H_0: $\tilde{\mu} = 24.2$ and H_A: $\tilde{\mu} > 24.2$

2. $\alpha = 0.01$

3. Reject H_0 if $z \geq 2.33$.

4. Without continuity correction $z = 2.333$ and with continuity correction $z = 2.17$.

5. Null hypothesis cannot be rejected.

18.9 The p-value is 0.01758.

18.11

1. $H_0 : \tilde{\mu}_D = 0$ and $H_A : \tilde{\mu}_D > 0$

2. $\alpha = 0.05$

3. Reject H_0 if $z \leq -1.645$

4. $z = \dfrac{14 - 19(0.5)}{\sqrt{19(0.5)(0.5)}} = -2.06$

5. The null hypothesis must be rejected

18.13

a. Reject H_0 if $T \leq 8$.

b. Reject H_0 if $T^- \leq 11$.

c. Reject H_0 if $T^+ \leq 11$.

18.15 **a.** Reject H_0 if $T \leq 7$.

      b. Reject H_0 if $T^- \leq 10$.

      c. Reject H_0 if $T^+ \leq 10$.

18.17 **a. 1.** H_0: $\mu = 45$ and H_A: $\mu < 45$

      2. $\alpha = 0.05$

      3. Reject H_0 if $T^+ \leq 21$.

      4. $T^+ = 18$

      5. The null hypothesis must be rejected.

      b. 1. H_0: $\mu = 0$ and H_A: $\mu \neq 0$

      2. $\alpha = 0.05$

      3. Reject H_0 if $T^+ \leq 17$

      4. $T^+ = 18$

      5. The null hypothesis cannot be rejected.

18.19 **1.** H_0: $\mu = 110$ and H_A: $\mu > 110$

      2. $\alpha = 0.01$

      3. Reject H_0 if $T^- \leq 33$

      4. $T^- = 18$

      5. The null hypothesis must be rejected.

18.21 **1.** H_0: $\mu_D = 0$ and H_A: $\mu_D < 0$

      2. $\alpha = 0.05$

      3. Reject H_0 if $T^+ \leq 26$.

      4. $T^+ = 5$

      5. The null hypothesis must be rejected.

18.23 **1.** and

      2. as in Exercise 18.21

      3. Reject H_0 if $z \leq -1.645$.

      4. Without continuity correction $z = \dfrac{5 - 52.5}{15.93} = -2.98$ and with continuity correction $z = -2.95$.

      5. The null hypothesis must be rejected.

18.25 **1.** and **2.** same as in Exercise 18.10

      3. Reject H_0 if $z \geq 1.645$.

      4. $z = 1.75$.

      5. The null hypothesis must be rejected.

18.31 **a.** Reject H_0 if $U_2 \leq 14$.

      b. Reject if $U \leq 11$.

      c. Reject if $U_1 \leq 14$.

18.33 **a.** Reject H_0 if $U_2 \leq 41$.

      b. Reject if $U \leq 36$.

      c. Reject if $U_1 \leq 41$.

18.35 **a.** Reject H_0 if $U_2 \leq 3$.

      b. Reject H_0 if $U_2 \leq 18$.

      c. Reject H_0 if $U_2 \leq 13$.

      d. Reject H_0 if $U_2 \leq 2$.

18.37 **1.** H_0: $\mu_1 = \mu_2$ and H_A: $\mu_1 \neq \mu_2$

      2. $\alpha = 0.05$

      3. Reject H_0 if $U \leq 37$.

      4. $W_1 = 188$, $W_2 = 112$, $U_1 = 110$, $U_2 = 34$, and $U = 34$.

      5. The null hypothesis must be rejected.

18.39 **1.** H_0: $\mu_1 = \mu_2$ and H_A: $\mu_1 \neq \mu_2$

      2. $\alpha = 0.05$

      3. Reject H_0 if $U \leq 49$.

      4. $W_1 = 208$, $W_2 = 170$, $U_1 = 88$, $U_2 = 92$, and $U = 88$.

      5. The null hypothesis cannot be rejected.

18.41   **1.** $H_0: \mu_1 = \mu_2$ and $H_A: \mu_1 < \mu_2$

2. $\alpha = 0.05$

3. Reject H_0 if $U_1 \leq 10$.

4. $W_1 = 26.5$ and $U_1 = 5.5$

5. The null hypothesis must be rejected.

18.43   **1.** $H_0: \mu_1 = \mu_2$ and $H_A: \mu_1 \neq \mu_2$

2. $\alpha = 0.05$

3. Reject H_0 if $z \leq -1.96$ or $z \geq 1.96$.

4. $z = \dfrac{24 - 45}{12.25} = -1.71$.

5. The null hypothesis cannot be rejected.

18.51   **1.** H_0: $\mu_1 = \mu_2 = \mu_3 = \mu_4$ and H_A: the μ's are not all equal.

        2. $\alpha = 0.05$

        3. Reject H_0 if $H \geq 7.815$.

        4. $R_1 = 53$, $R_2 = 68$, $R_3 = 30$, and $R_4 = 59$, and $H = 4.51$.

        5. The null hypothesis cannot be rejected.

18.53   **1.** H_0: $\mu_1 = \mu_2 = \mu_3$ and H_A: the μ's are not all equal.

        2. $\alpha = 0.01$

        3. Reject H_0 if $H \geq 9.210$.

        4. $R_1 = 121$, $R_2 = 144$, and $R_3 = 86$, and $H = 1.53$.

        5. The null hypothesis cannot be rejected.

18.55   **1.** H_0: Arrangement is random and H_A: arrangement is not random.

        2. $\alpha = 0.05$

        3. Reject H_0 if $u \leq 8$ or $u \geq 19$.

        4. $n_1 = 12$, $n_2 = 13$, and $u = 7$

        5. The null hypothesis must be rejected.

18.57   **1.** H_0: Arrangement is random and H_A: arrangement is not random.

        2. $\alpha = 0.01$

        3. Reject H_0 if $u \leq 8$ or $u \geq 23$.

        4. $n_1 = 15$, $n_2 = 14$, and $u = 20$.

        5. The null hypothesis cannot be rejected.

18.59   **1.** H_0: Arrangement is random and H_A: arrangement is not random.

        2. $\alpha = 0.05$

        3. Reject H_0 if $z \leq -1.96$ or $z \geq 1.96$.

        4. $z = \dfrac{7.5 - 10.92}{1.96} = -1.74$.

        5. The null hypothesis cannot be rejected.

18.61   **1.** H_0: Arrangement is random and H_A: arrangement is not random.

        2. $\alpha = 0.05$

        3. Reject H_0 if $z \leq -1.96$ or $z \geq 1.96$.

        4. $z = \dfrac{28 - 26.71}{3.40} = 0.38$.

        5. The null hypothesis cannot be rejected.

18.65   The median is 54.85 and the arrangement of values above and below the median is *aaaaabaabbbbbabbabaabaabaaababbbabbbabab abab*, so that $n_1 = 20$, $n_2 = 20$, and $u = 26$.

        1. H_0: Arrangement is random and H_A: arrangement is not random.

        2. $\alpha = 0.05$

        3. Reject H_0 if $z \leq -1.96$ or $z \geq 1.96$.

        4. $z = \dfrac{26 - 21}{3.12} = 1.60$.

        5. The null hypothesis cannot be rejected.

18.67   The arrangement is 140 and the arrangement of values above and below the median is *bbaabbaabbbbbbbaaaaababbbaaaaa*, so that $n_1 = 15$, $n_2 = 15$, and $u = 10$.

        1. H_0: Arrangement is random and H_A: arrangement is not random.

        2. $\alpha = 0.05$

        3. Reject H_0 if $z \leq -1.645$.

        4. $z = \dfrac{10.5 - 16}{2.69} = -2.04$.

        5. The null hypothesis must be rejected.

18.69 $\sum d^2 = 100$, so that

$$r_S = 1 - \frac{6 \cdot 100}{12 \cdot 143} = 0.65.$$

18.71 $z = 0.31\sqrt{49} = 2.17.$
Since $z = 2.17$ exceeds 1.96, the null
hypothesis must be rejected.

18.73 $r_s = 0.61$ for A and B, $r_s = -0.05$ for A and
C, and $r_s = -0.18$ for B and C.

  a. A and B are most alike

  b B and C are least alike.

Review Exercises for Chapters 15, 16, 17, and 18

R.169   They will belong to a group of families with a higher mean income, but no guaranteed increase.

R.171 $r = \dfrac{-103.8}{\sqrt{(312.1)(82.4)}} = -0.65.$

R.173 $n_1 = 23,\ n_2 = 7,$ and $u = 9$

       1. H_0: Arrangement is random and H_A: arrangement is not random.

       2. $\alpha = 0.01$

       3. Reject H_0 if $z \le -2.575$ or $z \ge 2.575$.

       4. $z = \dfrac{9 - 11.73}{1.90} = -1.44.$

       5. The null hypothesis cannot be rejected.

R.175 $r = \dfrac{-23.89}{\sqrt{(34,873.50)(0.0194)}} = -0.92$

R.177   **1.** H_0: $\mu_1 = \mu_2 = \mu_3 = \mu_4$ and H_A: the μ s are not all equal.

       2. $\alpha = 0.05$

       3. Reject H_0 if $F \ge 2.83$.

       4. $F = 3.48$

       5. The null hypothesis must be rejected.

R.179 $r = -1.03$, which is an impossible value.

R.181   **1.** and **2.** as in preceding exercise.

       3. Reject H_0 if $z \le -1.96$ or $z \ge 1.96$.

       4. $z = \dfrac{161 - 115.5}{28.77} = 1.58.$

       5. The null hypothesis cannot be rejected.

R.183   Since the males are all economists and the females are all statisticians, sex and field of specialization are confounded. There is no way in which we can distinguish between these two sources of variation on the basis of the experiment.

R.185   The three normal equations are
       $66.2 = 7a + 28c,\ -13.2 = 28b,$ and
       $165.8 = 28a + 196c,$ so that
       $a = 14.2,\ b = -0.471,$ and $c = -1.18.$ The equation of the parabola is $\hat{y} = 14.2 - 0.471x - 1.18x^2.$

R.187 The data yield $- - - + - - - + - - - - + - - +$, where $n = 16$ and $x = 4$.

1. H_0: $\rho = 0.50$ and H_A: $\rho < 0.50$.

2. $\alpha = 0.05$

3. x is the number of $+$ signs.

4. $x = 4$ and the p-value is 0.039.

5. The null hypothesis must be rejected.

R.189   **a.** Positive correlation

      b. no correlation

      c. positive correlation

      d. negative correlation

R.191 $r = \dfrac{1,727}{\sqrt{(20,456)(163.5)}} = 0.94$

R.193 $\left(\dfrac{-0.92}{0.41}\right)^2 = 5.04$

The second relationship is just about 5 times as strong as the first relationship.

R.195 $\bar{x}_1 = 11, \bar{x}_2 = 15, \bar{x}_3 = 10, \bar{x} = 12,$

$s_1^2 = \dfrac{26}{3}, s_2^2 = \dfrac{34}{3},$ and $s_3^2 = \dfrac{26}{3}.$

      a. $ns_{\bar{x}}^2 = 28$

$\dfrac{1}{3}(s_1^2 + s_2^2 + s_3^2) = \dfrac{86}{9},$ and $F = \dfrac{28}{\frac{86}{9}} = 2.93.$

      b. 1. H_0: $\mu_1 = \mu_2 = \mu_3$ and H_A: the μ's are not all equal.

        2. $\alpha = 0.01$

        3. Reject H_0 if $F \geq 8.02$

        4. $F = 2.93$

        5. The null hypothesis cannot be rejected.

R.197   **a.** It is a balanced incomplete block design because the seven department heads are not all serving together on a committee, but each department head serves together with each other department head on two committees.

      b. There are two solutions.

Griffith — Dramatics	Griffith — Dramatics
Anderson — Discipline	Anderson —
Evans — Tenure	Evans — Salaries
Fleming — Salaries	Fleming — Tenure

72

R.199　**a.** Reject H_0 if $U \leq 19$.

　　b. Reject H_0 if $U_1 \leq 23$.

　　c. Reject H_0 if $U_2 \leq 23$.

R.201　**1.** H_0: The row effects are all equal to zero; the column effects are all equal to zero; the treatment effects are all equal to zero.
　　　　H_A: The row effects are not all equal to zero. The column effects are not all equal to zero. The treatment effects are not all equal to zero.

　　2. $\alpha = 0.01$ for each test.

　　3. For each test, reject H_0 if $F \geq 5.41$

　　4. $F = 2.31$ for rows, $F = 8.24$ for columns, and $F = 31.28$ for treatments.

　　5. The null hypothesis for rows cannot be rejected. The null hypothesis for columns and treatments must both be rejected.

R.203　**1.** H_0: $\mu_1 = \mu_2$ and H_A: $\mu_1 \neq \mu_2$.

　　2. $\alpha = 0.05$

　　3. Reject H_0 if $z \leq -1.96$ or $z \geq 1.96$.

　　4. $z = 2.14$

　　5. The null hypothesis must be rejected.

R.205　**1.** H_0: $\tilde{\mu} = 169$ and $\tilde{H}_A$: $\mu \neq 169$.

　　2. $\alpha = 0.05$

　　3. Reject H_0 if $T \leq 11$.

　　4. $T = 11$ so that $T = 11$.

　　5. The null hypothesis must be rejected.

R.207　$\log \hat{y} = 1.45157 + 0.00698x$; $\log \hat{y} = 1.45157 + 0.00698(60)$
　　　　　　$= 1.8704$
　　$\hat{y} = 74.20$

R.211　The parabola provides an excellent fit for the given range of values of x, 50 to 65. However, the parabola will go up for greater values of x, and this does not make any sense for the given kind of data on price and demand.